Appendices to Accompany Fundamentals of Engineering Thermodynamics

MICHAEL J. MORAN
The Ohio State University

HOWARD N. SHAPIRO
Iowa State University

WILEY

Publisher	Don Fowley
Executive Editor	Linda Ratts
Editorial Assistant	Hope Ellis
Marketing Manager	Christopher Ruel
Design Director	Harry Nolan
Senior Content Manager	Kevin Holm
Senior Production Editor	Tim Lindner
Senior Designer	Madelyn Lesure
Senior Product Designer	Jenny Welter
Content Editor	Wendy Ashenberg
Photo Editor	Kathleen Pepper
Production Management Services	Aptara®, Inc.

Cover Photos: globe © DNY59 /iStockphoto, *left to right:* © shaunl/iStockphoto, © digitalskillet/ iStockphoto, © SelectStock/iStockphoto, © Mcelroyart/iStockphoto, © Aldo Murillo/iStockphoto, © technotr/iStockphoto, © digitalskillet/iStockphoto, © Shironosov/iStockphoto, © MichaelSvoboda/ iStockphoto, © gchutka/iStockphoto, © davidf/iStockphoto, © kupicoo/iStockphoto, © next999/ iStockphoto, Spine: © Estate of Stephen Laurence Strathdee/iStockphoto

 exxorian/iStockphoto

This book was typeset in 10/12 Times Ten Roman at Aptara®, Inc. and printed and bound by Quad/Graphics. The cover was printed by Quad/Graphics.

Founded in 1807, John Wiley & Sons, Inc. has been a valued source of knowledge and understanding for more than 200 years, helping people around the world meet their needs and fulfill their aspirations. Our company is built on a foundation of principles that include responsibility to the communities we serve and where we live and work. In 2008, we launched a Corporate Citizenship Initiative, a global effort to address the environmental, social, economic, and ethical challenges we face in our business. Among the issues we are addressing are carbon impact, paper specifications and procurement, ethical conduct within our business and among our vendors, and community and charitable support. For more information, please visit our website: *www.wiley.com/go/citizenship*.

The paper in this book was manufactured by a mill whose forest management programs include sustained yield-harvesting of its timberlands. Sustained yield harvesting principles ensure that the number of trees cut each year does not exceed the amount of new growth.

This book is printed on acid-free paper. ∞

ISBN   978-1-118-95721-9

Printed in the United States of America

V10002703_072718

Index to Tables in SI Units

TABLE A-1

Atomic or Molecular Weights and Critical Properties of Selected Elements and Compounds

Substance	Chemical Formula	M (kg/kmol)	T_c (K)	p_c (bar)	$Z_c = \dfrac{p_c v_c}{RT_c}$
Acetylene	C_2H_2	26.04	309	62.8	0.274
Air (equivalent)	—	28.97	133	37.7	0.284
Ammonia	NH_3	17.03	406	112.8	0.242
Argon	Ar	39.94	151	48.6	0.290
Benzene	C_6H_6	78.11	563	49.3	0.274
Butane	C_4H_{10}	58.12	425	38.0	0.274
Carbon	C	12.01	—	—	—
Carbon dioxide	CO_2	44.01	304	73.9	0.276
Carbon monoxide	CO	28.01	133	35.0	0.294
Copper	Cu	63.54	—	—	—
Ethane	C_2H_6	30.07	305	48.8	0.285
Ethanol	C_2H_5OH	46.07	516	63.8	0.249
Ethylene	C_2H_4	28.05	283	51.2	0.270
Helium	He	4.003	5.2	2.3	0.300
Hydrogen	H_2	2.016	33.2	13.0	0.304
Methane	CH_4	16.04	191	46.4	0.290
Methanol	CH_3OH	32.04	513	79.5	0.220
Nitrogen	N_2	28.01	126	33.9	0.291
Octane	C_8H_{18}	114.22	569	24.9	0.258
Oxygen	O_2	32.00	154	50.5	0.290
Propane	C_3H_8	44.09	370	42.7	0.276
Propylene	C_3H_6	42.08	365	46.2	0.276
Refrigerant 12	CCl_2F_2	120.92	385	41.2	0.278
Refrigerant 22	$CHClF_2$	86.48	369	49.8	0.267
Refrigerant 134a	CF_3CH_2F	102.03	374	40.7	0.260
Sulfur dioxide	SO_2	64.06	431	78.7	0.268
Water	H_2O	18.02	647.3	220.9	0.233

Sources: Adapted from *International Critical Tables* and L. C. Nelson and E. F. Obert, Generalized Compressibility Charts, *Chem. Eng., 61:* 203 (1954).

Sources for Tables A-2 through A-18.

Tables A-2 through A-6 are extracted from J. H. Keenan, F. G. Keyes, P. G. Hill, and J. G. Moore, *Steam Tables,* Wiley, New York, 1969.

Tables A-7 through A-9 are calculated based on equations from A. Kamei and S. W. Beyerlein, "A Fundamental Equation for Chlorodifluoromethane (R-22)," *Fluid Phase Equilibria,* Vol. 80, No. 11, 1992, pp. 71–86.

Tables A-10 through A-12 are calculated based on equations from D. P. Wilson and R. S. Basu, "Thermodynamic Properties of a New Stratospherically Safe Working Fluid — Refrigerant 134a," *ASHRAE Trans.,* Vol. 94, Pt. 2, 1988, pp. 2095–2118.

Tables A-13 through A-15 are calculated based on equations from L. Haar and J. S. Gallagher, "Thermodynamic Properties of Ammonia," *J. Phys. Chem. Reference Data,* Vol. 7, 1978, pp. 635–792.

Tables A-16 through A-18 are calculated based on B. A. Younglove and J. F. Ely, "Thermophysical Properties of Fluids. II. Methane, Ethane, Propane, Isobutane and Normal Butane," *J. Phys. Chem. Ref. Data,* Vol. 16, No. 4, 1987, pp. 577–598.

TABLE A-2

Properties of Saturated Water (Liquid–Vapor): Temperature Table

Pressure Conversions:
1 bar = 0.1 MPa
= 10^2 kPa

Temp. °C	Press. bar	Specific Volume m³/kg Sat. Liquid $v_f \times 10^3$	Specific Volume m³/kg Sat. Vapor v_g	Internal Energy kJ/kg Sat. Liquid u_f	Internal Energy kJ/kg Sat. Vapor u_g	Enthalpy kJ/kg Sat. Liquid h_f	Enthalpy kJ/kg Evap. h_{fg}	Enthalpy kJ/kg Sat. Vapor h_g	Entropy kJ/kg · K Sat. Liquid s_f	Entropy kJ/kg · K Sat. Vapor s_g	Temp. °C
.01	0.00611	1.0002	206.136	0.00	2375.3	0.01	2501.3	2501.4	0.0000	9.1562	.01
4	0.00813	1.0001	157.232	16.77	2380.9	16.78	2491.9	2508.7	0.0610	9.0514	4
5	0.00872	1.0001	147.120	20.97	2382.3	20.98	2489.6	2510.6	0.0761	9.0257	5
6	0.00935	1.0001	137.734	25.19	2383.6	25.20	2487.2	2512.4	0.0912	9.0003	6
8	0.01072	1.0002	120.917	33.59	2386.4	33.60	2482.5	2516.1	0.1212	8.9501	8
10	0.01228	1.0004	106.379	42.00	2389.2	42.01	2477.7	2519.8	0.1510	8.9008	10
11	0.01312	1.0004	99.857	46.20	2390.5	46.20	2475.4	2521.6	0.1658	8.8765	11
12	0.01402	1.0005	93.784	50.41	2391.9	50.41	2473.0	2523.4	0.1806	8.8524	12
13	0.01497	1.0007	88.124	54.60	2393.3	54.60	2470.7	2525.3	0.1953	8.8285	13
14	0.01598	1.0008	82.848	58.79	2394.7	58.80	2468.3	2527.1	0.2099	8.8048	14
15	0.01705	1.0009	77.926	62.99	2396.1	62.99	2465.9	2528.9	0.2245	8.7814	15
16	0.01818	1.0011	73.333	67.18	2397.4	67.19	2463.6	2530.8	0.2390	8.7582	16
17	0.01938	1.0012	69.044	71.38	2398.8	71.38	2461.2	2532.6	0.2535	8.7351	17
18	0.02064	1.0014	65.038	75.57	2400.2	75.58	2458.8	2534.4	0.2679	8.7123	18
19	0.02198	1.0016	61.293	79.76	2401.6	79.77	2456.5	2536.2	0.2823	8.6897	19
20	0.02339	1.0018	57.791	83.95	2402.9	83.96	2454.1	2538.1	0.2966	8.6672	20
21	0.02487	1.0020	54.514	88.14	2404.3	88.14	2451.8	2539.9	0.3109	8.6450	21
22	0.02645	1.0022	51.447	92.32	2405.7	92.33	2449.4	2541.7	0.3251	8.6229	22
23	0.02810	1.0024	48.574	96.51	2407.0	96.52	2447.0	2543.5	0.3393	8.6011	23
24	0.02985	1.0027	45.883	100.70	2408.4	100.70	2444.7	2545.4	0.3534	8.5794	24
25	0.03169	1.0029	43.360	104.88	2409.8	104.89	2442.3	2547.2	0.3674	8.5580	25
26	0.03363	1.0032	40.994	109.06	2411.1	109.07	2439.9	2549.0	0.3814	8.5367	26
27	0.03567	1.0035	38.774	113.25	2412.5	113.25	2437.6	2550.8	0.3954	8.5156	27
28	0.03782	1.0037	36.690	117.42	2413.9	117.43	2435.2	2552.6	0.4093	8.4946	28
29	0.04008	1.0040	34.733	121.60	2415.2	121.61	2432.8	2554.5	0.4231	8.4739	29
30	0.04246	1.0043	32.894	125.78	2416.6	125.79	2430.5	2556.3	0.4369	8.4533	30
31	0.04496	1.0046	31.165	129.96	2418.0	129.97	2428.1	2558.1	0.4507	8.4329	31
32	0.04759	1.0050	29.540	134.14	2419.3	134.15	2425.7	2559.9	0.4644	8.4127	32
33	0.05034	1.0053	28.011	138.32	2420.7	138.33	2423.4	2561.7	0.4781	8.3927	33
34	0.05324	1.0056	26.571	142.50	2422.0	142.50	2421.0	2563.5	0.4917	8.3728	34
35	0.05628	1.0060	25.216	146.67	2423.4	146.68	2418.6	2565.3	0.5053	8.3531	35
36	0.05947	1.0063	23.940	150.85	2424.7	150.86	2416.2	2567.1	0.5188	8.3336	36
38	0.06632	1.0071	21.602	159.20	2427.4	159.21	2411.5	2570.7	0.5458	8.2950	38
40	0.07384	1.0078	19.523	167.56	2430.1	167.57	2406.7	2574.3	0.5725	8.2570	40
45	0.09593	1.0099	15.258	188.44	2436.8	188.45	2394.8	2583.2	0.6387	8.1648	45

H₂O

v_f = (table value)/1000

TABLE A-2

H₂O

(Continued)

Pressure Conversions:
1 bar = 0.1 MPa
= 10² kPa

v_f = (table value)/1000

Temp. °C	Press. bar	Specific Volume m³/kg		Internal Energy kJ/kg		Enthalpy kJ/kg			Entropy kJ/kg · K		Temp. °C
		Sat. Liquid $v_f \times 10^3$	Sat. Vapor v_g	Sat. Liquid u_f	Sat. Vapor u_g	Sat. Liquid h_f	Evap. h_{fg}	Sat. Vapor h_g	Sat. Liquid s_f	Sat. Vapor s_g	
50	.1235	1.0121	12.032	209.32	2443.5	209.33	2382.7	2592.1	.7038	8.0763	50
55	.1576	1.0146	9.568	230.21	2450.1	230.23	2370.7	2600.9	.7679	7.9913	55
60	.1994	1.0172	7.671	251.11	2456.6	251.13	2358.5	2609.6	.8312	7.9096	60
65	.2503	1.0199	6.197	272.02	2463.1	272.06	2346.2	2618.3	.8935	7.8310	65
70	.3119	1.0228	5.042	292.95	2469.6	292.98	2333.8	2626.8	.9549	7.7553	70
75	.3858	1.0259	4.131	313.90	2475.9	313.93	2321.4	2635.3	1.0155	7.6824	75
80	.4739	1.0291	3.407	334.86	2482.2	334.91	2308.8	2643.7	1.0753	7.6122	80
85	.5783	1.0325	2.828	355.84	2488.4	355.90	2296.0	2651.9	1.1343	7.5445	85
90	.7014	1.0360	2.361	376.85	2494.5	376.92	2283.2	2660.1	1.1925	7.4791	90
95	.8455	1.0397	1.982	397.88	2500.6	397.96	2270.2	2668.1	1.2500	7.4159	95
100	1.014	1.0435	1.673	418.94	2506.5	419.04	2257.0	2676.1	1.3069	7.3549	100
110	1.433	1.0516	1.210	461.14	2518.1	461.30	2230.2	2691.5	1.4185	7.2387	110
120	1.985	1.0603	0.8919	503.50	2529.3	503.71	2202.6	2706.3	1.5276	7.1296	120
130	2.701	1.0697	0.6685	546.02	2539.9	546.31	2174.2	2720.5	1.6344	7.0269	130
140	3.613	1.0797	0.5089	588.74	2550.0	589.13	2144.7	2733.9	1.7391	6.9299	140
150	4.758	1.0905	0.3928	631.68	2559.5	632.20	2114.3	2746.5	1.8418	6.8379	150
160	6.178	1.1020	0.3071	674.86	2568.4	675.55	2082.6	2758.1	1.9427	6.7502	160
170	7.917	1.1143	0.2428	718.33	2576.5	719.21	2049.5	2768.7	2.0419	6.6663	170
180	10.02	1.1274	0.1941	762.09	2583.7	763.22	2015.0	2778.2	2.1396	6.5857	180
190	12.54	1.1414	0.1565	806.19	2590.0	807.62	1978.8	2786.4	2.2359	6.5079	190
200	15.54	1.1565	0.1274	850.65	2595.3	852.45	1940.7	2793.2	2.3309	6.4323	200
210	19.06	1.1726	0.1044	895.53	2599.5	897.76	1900.7	2798.5	2.4248	6.3585	210
220	23.18	1.1900	0.08619	940.87	2602.4	943.62	1858.5	2802.1	2.5178	6.2861	220
230	27.95	1.2088	0.07158	986.74	2603.9	990.12	1813.8	2804.0	2.6099	6.2146	230
240	33.44	1.2291	0.05976	1033.2	2604.0	1037.3	1766.5	2803.8	2.7015	6.1437	240
250	39.73	1.2512	0.05013	1080.4	2602.4	1085.4	1716.2	2801.5	2.7927	6.0730	250
260	46.88	1.2755	0.04221	1128.4	2599.0	1134.4	1662.5	2796.6	2.8838	6.0019	260
270	54.99	1.3023	0.03564	1177.4	2593.7	1184.5	1605.2	2789.7	2.9751	5.9301	270
280	64.12	1.3321	0.03017	1227.5	2586.1	1236.0	1543.6	2779.6	3.0668	5.8571	280
290	74.36	1.3656	0.02557	1278.9	2576.0	1289.1	1477.1	2766.2	3.1594	5.7821	290
300	85.81	1.4036	0.02167	1332.0	2563.0	1344.0	1404.9	2749.0	3.2534	5.7045	300
320	112.7	1.4988	0.01549	1444.6	2525.5	1461.5	1238.6	2700.1	3.4480	5.5362	320
340	145.9	1.6379	0.01080	1570.3	2464.6	1594.2	1027.9	2622.0	3.6594	5.3357	340
360	186.5	1.8925	0.006945	1725.2	2351.5	1760.5	720.5	2481.0	3.9147	5.0526	360
374.14	220.9	3.155	0.003155	2029.6	2029.6	2099.3	0	2099.3	4.4298	4.4298	374.14

TABLE A-3

Properties of Saturated Water (Liquid–Vapor): Pressure Table

Pressure Conversions:
1 bar = 0.1 MPa
= 10^2 kPa

v_f = (table value)/1000

H₂O

Press. bar	Temp. °C	Specific Volume m³/kg		Internal Energy kJ/kg		Enthalpy kJ/kg			Entropy kJ/kg · K		Press. bar
		Sat. Liquid $v_f \times 10^3$	Sat. Vapor v_g	Sat. Liquid u_f	Sat. Vapor u_g	Sat. Liquid h_f	Evap. h_{fg}	Sat. Vapor h_g	Sat. Liquid s_f	Sat. Vapor s_g	
0.04	28.96	1.0040	34.800	121.45	2415.2	121.46	2432.9	2554.4	0.4226	8.4746	0.04
0.06	36.16	1.0064	23.739	151.53	2425.0	151.53	2415.9	2567.4	0.5210	8.3304	0.06
0.08	41.51	1.0084	18.103	173.87	2432.2	173.88	2403.1	2577.0	0.5926	8.2287	0.08
0.10	45.81	1.0102	14.674	191.82	2437.9	191.83	2392.8	2584.7	0.6493	8.1502	0.10
0.20	60.06	1.0172	7.649	251.38	2456.7	251.40	2358.3	2609.7	0.8320	7.9085	0.20
0.30	69.10	1.0223	5.229	289.20	2468.4	289.23	2336.1	2625.3	0.9439	7.7686	0.30
0.40	75.87	1.0265	3.993	317.53	2477.0	317.58	2319.2	2636.8	1.0259	7.6700	0.40
0.50	81.33	1.0300	3.240	340.44	2483.9	340.49	2305.4	2645.9	1.0910	7.5939	0.50
0.60	85.94	1.0331	2.732	359.79	2489.6	359.86	2293.6	2653.5	1.1453	7.5320	0.60
0.70	89.95	1.0360	2.365	376.63	2494.5	376.70	2283.3	2660.0	1.1919	7.4797	0.70
0.80	93.50	1.0380	2.087	391.58	2498.8	391.66	2274.1	2665.8	1.2329	7.4346	0.80
0.90	96.71	1.0410	1.869	405.06	2502.6	405.15	2265.7	2670.9	1.2695	7.3949	0.90
1.00	99.63	1.0432	1.694	417.36	2506.1	417.46	2258.0	2675.5	1.3026	7.3594	1.00
1.50	111.4	1.0528	1.159	466.94	2519.7	467.11	2226.5	2693.6	1.4336	7.2233	1.50
2.00	120.2	1.0605	0.8857	504.49	2529.5	504.70	2201.9	2706.7	1.5301	7.1271	2.00
2.50	127.4	1.0672	0.7187	535.10	2537.2	535.37	2181.5	2716.9	1.6072	7.0527	2.50
3.00	133.6	1.0732	0.6058	561.15	2543.6	561.47	2163.8	2725.3	1.6718	6.9919	3.00
3.50	138.9	1.0786	0.5243	583.95	2546.9	584.33	2148.1	2732.4	1.7275	6.9405	3.50
4.00	143.6	1.0836	0.4625	604.31	2553.6	604.74	2133.8	2738.6	1.7766	6.8959	4.00
4.50	147.9	1.0882	0.4140	622.25	2557.6	623.25	2120.7	2743.9	1.8207	6.8565	4.50
5.00	151.9	1.0926	0.3749	639.68	2561.2	640.23	2108.5	2748.7	1.8607	6.8212	5.00
6.00	158.9	1.1006	0.3157	669.90	2567.4	670.56	2086.3	2756.8	1.9312	6.7600	6.00
7.00	165.0	1.1080	0.2729	696.44	2572.5	697.22	2066.3	2763.5	1.9922	6.7080	7.00
8.00	170.4	1.1148	0.2404	720.22	2576.8	721.11	2048.0	2769.1	2.0462	6.6628	8.00
9.00	175.4	1.1212	0.2150	741.83	2580.5	742.83	2031.1	2773.9	2.0946	6.6226	9.00
10.0	179.9	1.1273	0.1944	761.68	2583.6	762.81	2015.3	2778.1	2.1387	6.5863	10.0
15.0	198.3	1.1539	0.1318	843.16	2594.5	844.84	1947.3	2792.2	2.3150	6.4448	15.0
20.0	212.4	1.1767	0.09963	906.44	2600.3	908.79	1890.7	2799.5	2.4474	6.3409	20.0
25.0	224.0	1.1973	0.07998	959.11	2603.1	962.11	1841.0	2803.1	2.5547	6.2575	25.0
30.0	233.9	1.2165	0.06668	1004.8	2604.1	1008.4	1795.7	2804.2	2.6457	6.1869	30.0
35.0	242.6	1.2347	0.05707	1045.4	2603.7	1049.8	1753.7	2803.4	2.7253	6.1253	35.0
40.0	250.4	1.2522	0.04978	1082.3	2602.3	1087.3	1714.1	2801.4	2.7964	6.0701	40.0
45.0	257.5	1.2692	0.04406	1116.2	2600.1	1121.9	1676.4	2798.3	2.8610	6.0199	45.0
50.0	264.0	1.2859	0.03944	1147.8	2597.1	1154.2	1640.1	2794.3	2.9202	5.9734	50.0
60.0	275.6	1.3187	0.03244	1205.4	2589.7	1213.4	1571.0	2784.3	3.0267	5.8892	60.0
70.0	285.9	1.3513	0.02737	1257.6	2580.5	1267.0	1505.1	2772.1	3.1211	5.8133	70.0
80.0	295.1	1.3842	0.02352	1305.6	2569.8	1316.6	1441.3	2758.0	3.2068	5.7432	80.0
90.0	303.4	1.4178	0.02048	1350.5	2557.8	1363.3	1378.9	2742.1	3.2858	5.6772	90.0
100.	311.1	1.4524	0.01803	1393.0	2544.4	1407.6	1317.1	2724.7	3.3596	5.6141	100.
110.	318.2	1.4886	0.01599	1433.7	2529.8	1450.1	1255.5	2705.6	3.4295	5.5527	110.

TABLE A-3

H₂O

(Continued)

Pressure Conversions:
1 bar = 0.1 MPa
= 10² kPa

Press. bar	Temp. °C	Specific Volume m³/kg		Internal Energy kJ/kg		Enthalpy kJ/kg			Entropy kJ/kg · K		Press. bar
		Sat. Liquid $v_f \times 10^3$	Sat. Vapor v_g	Sat. Liquid u_f	Sat. Vapor u_g	Sat. Liquid h_f	Evap. h_{fg}	Sat. Vapor h_g	Sat. Liquid s_f	Sat. Vapor s_g	
120.	324.8	1.5267	0.01426	1473.0	2513.7	1491.3	1193.6	2684.9	3.4962	5.4924	120.
130.	330.9	1.5671	0.01278	1511.1	2496.1	1531.5	1130.7	2662.2	3.5606	5.4323	130.
140.	336.8	1.6107	0.01149	1548.6	2476.8	1571.1	1066.5	2637.6	3.6232	5.3717	140.
150.	342.2	1.6581	0.01034	1585.6	2455.5	1610.5	1000.0	2610.5	3.6848	5.3098	150.
160.	347.4	1.7107	0.009306	1622.7	2431.7	1650.1	930.6	2580.6	3.7461	5.2455	160.
170.	352.4	1.7702	0.008364	1660.2	2405.0	1690.3	856.9	2547.2	3.8079	5.1777	170.
180.	357.1	1.8397	0.007489	1698.9	2374.3	1732.0	777.1	2509.1	3.8715	5.1044	180.
190.	361.5	1.9243	0.006657	1739.9	2338.1	1776.5	688.0	2464.5	3.9388	5.0228	190.
200.	365.8	2.036	0.005834	1785.6	2293.0	1826.3	583.4	2409.7	4.0139	4.9269	200.
220.9	374.1	3.155	0.003155	2029.6	2029.6	2099.3	0	2099.3	4.4298	4.4298	220.9

v_f = (table value)/1000

TABLE A-4

Properties of Superheated Water Vapor

H₂O

Pressure Conversions:
1 bar = 0.1 MPa
= 10² kPa

T °C	v m³/kg	u kJ/kg	h kJ/kg	s kJ/kg·K	v m³/kg	u kJ/kg	h kJ/kg	s kJ/kg·K
	p = 0.06 bar = 0.006 MPa (T_sat = 36.16°C)				p = 0.35 bar = 0.035 MPa (T_sat = 72.69°C)			
Sat.	23.739	2425.0	2567.4	8.3304	4.526	2473.0	2631.4	7.7158
80	27.132	2487.3	2650.1	8.5804	4.625	2483.7	2645.6	7.7564
120	30.219	2544.7	2726.0	8.7840	5.163	2542.4	2723.1	7.9644
160	33.302	2602.7	2802.5	8.9693	5.696	2601.2	2800.6	8.1519
200	36.383	2661.4	2879.7	9.1398	6.228	2660.4	2878.4	8.3237
240	39.462	2721.0	2957.8	9.2982	6.758	2720.3	2956.8	8.4828
280	42.540	2781.5	3036.8	9.4464	7.287	2780.9	3036.0	8.6314
320	45.618	2843.0	3116.7	9.5859	7.815	2842.5	3116.1	8.7712
360	48.696	2905.5	3197.7	9.7180	8.344	2905.1	3197.1	8.9034
400	51.774	2969.0	3279.6	9.8435	8.872	2968.6	3279.2	9.0291
440	54.851	3033.5	3362.6	9.9633	9.400	3033.2	3362.2	9.1490
500	59.467	3132.3	3489.1	10.1336	10.192	3132.1	3488.8	9.3194
	p = 0.70 bar = 0.07 MPa (T_sat = 89.95°C)				p = 1.0 bar = 0.10 MPa (T_sat = 99.63°C)			
Sat.	2.365	2494.5	2660.0	7.4797	1.694	2506.1	2675.5	7.3594
100	2.434	2509.7	2680.0	7.5341	1.696	2506.7	2676.2	7.3614
120	2.571	2539.7	2719.6	7.6375	1.793	2537.3	2716.6	7.4668
160	2.841	2599.4	2798.2	7.8279	1.984	2597.8	2796.2	7.6597
200	3.108	2659.1	2876.7	8.0012	2.172	2658.1	2875.3	7.8343
240	3.374	2719.3	2955.5	8.1611	2.359	2718.5	2954.5	7.9949
280	3.640	2780.2	3035.0	8.3162	2.546	2779.6	3034.2	8.1445
320	3.905	2842.0	3115.3	8.4504	2.732	2841.5	3114.6	8.2849
360	4.170	2904.6	3196.5	8.5828	2.917	2904.2	3195.9	8.4175
400	4.434	2968.2	3278.6	8.7086	3.103	2967.9	3278.2	8.5435
440	4.698	3032.9	3361.8	8.8286	3.288	3032.6	3361.4	8.6636
500	5.095	3131.8	3488.5	8.9991	3.565	3131.6	3488.1	8.8342
	p = 1.5 bar = 0.15 MPa (T_sat = 111.37°C)				p = 3.0 bar = 0.30 MPa (T_sat = 133.55°C)			
Sat.	1.159	2519.7	2693.6	7.2233	0.606	2543.6	2725.3	6.9919
120	1.188	2533.3	2711.4	7.2693				
160	1.317	2595.2	2792.8	7.4665	0.651	2587.1	2782.3	7.1276
200	1.444	2656.2	2872.9	7.6433	0.716	2650.7	2865.5	7.3115
240	1.570	2717.2	2952.7	7.8052	0.781	2713.1	2947.3	7.4774
280	1.695	2778.6	3032.8	7.9555	0.844	2775.4	3028.6	7.6299
320	1.819	2840.6	3113.5	8.0964	0.907	2838.1	3110.1	7.7722
360	1.943	2903.5	3195.0	8.2293	0.969	2901.4	3192.2	7.9061
400	2.067	2967.3	3277.4	8.3555	1.032	2965.6	3275.0	8.0330
440	2.191	3032.1	3360.7	8.4757	1.094	3030.6	3358.7	8.1538
500	2.376	3131.2	3487.6	8.6466	1.187	3130.0	3486.0	8.3251
600	2.685	3301.7	3704.3	8.9101	1.341	3300.8	3703.2	8.5892

H₂O

Pressure Conversions:
1 bar = 0.1 MPa
= 10² kPa

TABLE A-4

(Continued)

T	v	u	h	s	v	u	h	s
°C	m³/kg	kJ/kg	kJ/kg	kJ/kg · K	m³/kg	kJ/kg	kJ/kg	kJ/kg · K
	p = 5.0 bar = 0.50 MPa (T_sat = 151.86°C)				p = 7.0 bar = 0.70 MPa (T_sat = 164.97°C)			
Sat.	0.3749	2561.2	2748.7	6.8213	0.2729	2572.5	2763.5	6.7080
180	0.4045	2609.7	2812.0	6.9656	0.2847	2599.8	2799.1	6.7880
200	0.4249	2642.9	2855.4	7.0592	0.2999	2634.8	2844.8	6.8865
240	0.4646	2707.6	2939.9	7.2307	0.3292	2701.8	2932.2	7.0641
280	0.5034	2771.2	3022.9	7.3865	0.3574	2766.9	3017.1	7.2233
320	0.5416	2834.7	3105.6	7.5308	0.3852	2831.3	3100.9	7.3697
360	0.5796	2898.7	3188.4	7.6660	0.4126	2895.8	3184.7	7.5063
400	0.6173	2963.2	3271.9	7.7938	0.4397	2960.9	3268.7	7.6350
440	0.6548	3028.6	3356.0	7.9152	0.4667	3026.6	3353.3	7.7571
500	0.7109	3128.4	3483.9	8.0873	0.5070	3126.8	3481.7	7.9299
600	0.8041	3299.6	3701.7	8.3522	0.5738	3298.5	3700.2	8.1956
700	0.8969	3477.5	3925.9	8.5952	0.6403	3476.6	3924.8	8.4391
	p = 10.0 bar = 1.0 MPa (T_sat = 179.91°C)				p = 15.0 bar = 1.5 MPa (T_sat = 198.32°C)			
Sat.	0.1944	2583.6	2778.1	6.5865	0.1318	2594.5	2792.2	6.4448
200	0.2060	2621.9	2827.9	6.6940	0.1325	2598.1	2796.8	6.4546
240	0.2275	2692.9	2920.4	6.8817	0.1483	2676.9	2899.3	6.6628
280	0.2480	2760.2	3008.2	7.0465	0.1627	2748.6	2992.7	6.8381
320	0.2678	2826.1	3093.9	7.1962	0.1765	2817.1	3081.9	6.9938
360	0.2873	2891.6	3178.9	7.3349	0.1899	2884.4	3169.2	7.1363
400	0.3066	2957.3	3263.9	7.4651	0.2030	2951.3	3255.8	7.2690
440	0.3257	3023.6	3349.3	7.5883	0.2160	3018.5	3342.5	7.3940
500	0.3541	3124.4	3478.5	7.7622	0.2352	3120.3	3473.1	7.5698
540	0.3729	3192.6	3565.6	7.8720	0.2478	3189.1	3560.9	7.6805
600	0.4011	3296.8	3697.9	8.0290	0.2668	3293.9	3694.0	7.8385
640	0.4198	3367.4	3787.2	8.1290	0.2793	3364.8	3783.8	7.9391
	p = 20.0 bar = 2.0 MPa (T_sat = 212.42°C)				p = 30.0 bar = 3.0 MPa (T_sat = 233.90°C)			
Sat.	0.0996	2600.3	2799.5	6.3409	0.0667	2604.1	2804.2	6.1869
240	0.1085	2659.6	2876.5	6.4952	0.0682	2619.7	2824.3	6.2265
280	0.1200	2736.4	2976.4	6.6828	0.0771	2709.9	2941.3	6.4462
320	0.1308	2807.9	3069.5	6.8452	0.0850	2788.4	3043.4	6.6245
360	0.1411	2877.0	3159.3	6.9917	0.0923	2861.7	3138.7	6.7801
400	0.1512	2945.2	3247.6	7.1271	0.0994	2932.8	3230.9	6.9212
440	0.1611	3013.4	3335.5	7.2540	0.1062	3002.9	3321.5	7.0520
500	0.1757	3116.2	3467.6	7.4317	0.1162	3108.0	3456.5	7.2338
540	0.1853	3185.6	3556.1	7.5434	0.1227	3178.4	3546.6	7.3474
600	0.1996	3290.9	3690.1	7.7024	0.1324	3285.0	3682.3	7.5085
640	0.2091	3362.2	3780.4	7.8035	0.1388	3357.0	3773.5	7.6106
700	0.2232	3470.9	3917.4	7.9487	0.1484	3466.5	3911.7	7.7571

TABLE A-4

(Continued)

H₂O

Pressure Conversions:
1 bar = 0.1 MPa
= 10² kPa

T °C	v m³/kg	u kJ/kg	h kJ/kg	s kJ/kg·K	v m³/kg	u kJ/kg	h kJ/kg	s kJ/kg·K
	p = 40 bar = 4.0 MPa (T_sat = 250.4°C)				p = 60 bar = 6.0 MPa (T_sat = 275.64°C)			
Sat.	0.04978	2602.3	2801.4	6.0701	0.03244	2589.7	2784.3	5.8892
280	0.05546	2680.0	2901.8	6.2568	0.03317	2605.2	2804.2	5.9252
320	0.06199	2767.4	3015.4	6.4553	0.03876	2720.0	2952.6	6.1846
360	0.06788	2845.7	3117.2	6.6215	0.04331	2811.2	3071.1	6.3782
400	0.07341	2919.9	3213.6	6.7690	0.04739	2892.9	3177.2	6.5408
440	0.07872	2992.2	3307.1	6.9041	0.05122	2970.0	3277.3	6.6853
500	0.08643	3099.5	3445.3	7.0901	0.05665	3082.2	3422.2	6.8803
540	0.09145	3171.1	3536.9	7.2056	0.06015	3156.1	3517.0	6.9999
600	0.09885	3279.1	3674.4	7.3688	0.06525	3266.9	3658.4	7.1677
640	0.1037	3351.8	3766.6	7.4720	0.06859	3341.0	3752.6	7.2731
700	0.1110	3462.1	3905.9	7.6198	0.07352	3453.1	3894.1	7.4234
740	0.1157	3536.6	3999.6	7.7141	0.07677	3528.3	3989.2	7.5190
	p = 80 bar = 8.0 MPa (T_sat = 295.06°C)				p = 100 bar = 10.0 MPa (T_sat = 311.06°C)			
Sat.	0.02352	2569.8	2758.0	5.7432	0.01803	2544.4	2724.7	5.6141
320	0.02682	2662.7	2877.2	5.9489	0.01925	2588.8	2781.3	5.7103
360	0.03089	2772.7	3019.8	6.1819	0.02331	2729.1	2962.1	6.0060
400	0.03432	2863.8	3138.3	6.3634	0.02641	2832.4	3096.5	6.2120
440	0.03742	2946.7	3246.1	6.5190	0.02911	2922.1	3213.2	6.3805
480	0.04034	3025.7	3348.4	6.6586	0.03160	3005.4	3321.4	6.5282
520	0.04313	3102.7	3447.7	6.7871	0.03394	3085.6	3425.1	6.6622
560	0.04582	3178.7	3545.3	6.9072	0.03619	3164.1	3526.0	6.7864
600	0.04845	3254.4	3642.0	7.0206	0.03837	3241.7	3625.3	6.9029
640	0.05102	3330.1	3738.3	7.1283	0.04048	3318.9	3723.7	7.0131
700	0.05481	3443.9	3882.4	7.2812	0.04358	3434.7	3870.5	7.1687
740	0.05729	3520.4	3978.7	7.3782	0.04560	3512.1	3968.1	7.2670
	p = 120 bar = 12.0 MPa (T_sat = 324.75°C)				p = 140 bar = 14.0 MPa (T_sat = 336.75°C)			
Sat.	0.01426	2513.7	2684.9	5.4924	0.01149	2476.8	2637.6	5.3717
360	0.01811	2678.4	2895.7	5.8361	0.01422	2617.4	2816.5	5.6602
400	0.02108	2798.3	3051.3	6.0747	0.01722	2760.9	3001.9	5.9448
440	0.02355	2896.1	3178.7	6.2586	0.01954	2868.6	3142.2	6.1474
480	0.02576	2984.4	3293.5	6.4154	0.02157	2962.5	3264.5	6.3143
520	0.02781	3068.0	3401.8	6.5555	0.02343	3049.8	3377.8	6.4610
560	0.02977	3149.0	3506.2	6.6840	0.02517	3133.6	3486.0	6.5941
600	0.03164	3228.7	3608.3	6.8037	0.02683	3215.4	3591.1	6.7172
640	0.03345	3307.5	3709.0	6.9164	0.02843	3296.0	3694.1	6.8326
700	0.03610	3425.2	3858.4	7.0749	0.03075	3415.7	3846.2	6.9939
740	0.03781	3503.7	3957.4	7.1746	0.03225	3495.2	3946.7	7.0952

H₂O

Pressure Conversions:
1 bar = 0.1 MPa
     = 10² kPa

TABLE A-4
(Continued)

T °C	v m³/kg	u kJ/kg	h kJ/kg	s kJ/kg·K	v m³/kg	u kJ/kg	h kJ/kg	s kJ/kg·K
	p = 160 bar = 16.0 MPa (Tsat = 347.44°C)				p = 180 bar = 18.0 MPa (Tsat = 357.06°C)			
Sat.	0.00931	2431.7	2580.6	5.2455	0.00749	2374.3	2509.1	5.1044
360	0.01105	2539.0	2715.8	5.4614	0.00809	2418.9	2564.5	5.1922
400	0.01426	2719.4	2947.6	5.8175	0.01190	2672.8	2887.0	5.6887
440	0.01652	2839.4	3103.7	6.0429	0.01414	2808.2	3062.8	5.9428
480	0.01842	2939.7	3234.4	6.2215	0.01596	2915.9	3203.2	6.1345
520	0.02013	3031.1	3353.3	6.3752	0.01757	3011.8	3378.0	6.2960
560	0.02172	3117.8	3465.4	6.5132	0.01904	3101.7	3444.4	6.4392
600	0.02323	3201.8	3573.5	6.6399	0.02042	3188.0	3555.6	6.5696
640	0.02467	3284.2	3678.9	6.7580	0.02174	3272.3	3663.6	6.6905
700	0.02674	3406.0	3833.9	6.9224	0.02362	3396.3	3821.5	6.8580
740	0.02808	3486.7	3935.9	7.0251	0.02483	3478.0	3925.0	6.9623
	p = 200 bar = 20.0 MPa (Tsat = 365.81°C)				p = 240 bar = 24.0 MPa			
Sat.	0.00583	2293.0	2409.7	4.9269				
400	0.00994	2619.3	2818.1	5.5540	0.00673	2477.8	2639.4	5.2393
440	0.01222	2774.9	3019.4	5.8450	0.00929	2700.6	2923.4	5.6506
480	0.01399	2891.2	3170.8	6.0518	0.01100	2838.3	3102.3	5.8950
520	0.01551	2992.0	3302.2	6.2218	0.01241	2950.5	3248.5	6.0842
560	0.01689	3085.2	3423.0	6.3705	0.01366	3051.1	3379.0	6.2448
600	0.01818	3174.0	3537.6	6.5048	0.01481	3145.2	3500.7	6.3875
640	0.01940	3260.2	3648.1	6.6286	0.01588	3235.5	3616.7	6.5174
700	0.02113	3386.4	3809.0	6.7993	0.01739	3366.4	3783.8	6.6947
740	0.02224	3469.3	3914.1	6.9052	0.01835	3451.7	3892.1	6.8038
800	0.02385	3592.7	4069.7	7.0544	0.01974	3578.0	4051.6	6.9567
	p = 280 bar = 28.0 MPa				p = 320 bar = 32.0 MPa			
400	0.00383	2223.5	2330.7	4.7494	0.00236	1980.4	2055.9	4.3239
440	0.00712	2613.2	2812.6	5.4494	0.00544	2509.0	2683.0	5.2327
480	0.00885	2780.8	3028.5	5.7446	0.00722	2718.1	2949.2	5.5968
520	0.01020	2906.8	3192.3	5.9566	0.00853	2860.7	3133.7	5.8357
560	0.01136	3015.7	3333.7	6.1307	0.00963	2979.0	3287.2	6.0246
600	0.01241	3115.6	3463.0	6.2823	0.01061	3085.3	3424.6	6.1858
640	0.01338	3210.3	3584.8	6.4187	0.01150	3184.5	3552.5	6.3290
700	0.01473	3346.1	3758.4	6.6029	0.01273	3325.4	3732.8	6.5203
740	0.01558	3433.9	3870.0	6.7153	0.01350	3415.9	3847.8	6.6361
800	0.01680	3563.1	4033.4	6.8720	0.01460	3548.0	4015.1	6.7966
900	0.01873	3774.3	4298.8	7.1084	0.01633	3762.7	4285.1	7.0372

TABLE A-5

Properties of Compressed Liquid Water

Pressure Conversions:
1 bar = 0.1 MPa
= 10^2 kPa

H₂O

T °C	$v \times 10^3$ m³/kg	u kJ/kg	h kJ/kg	s kJ/kg·K	$v \times 10^3$ m³/kg	u kJ/kg	h kJ/kg	s kJ/kg·K
	p = 25 bar = 2.5 MPa				**p = 50 bar = 5.0 MPa**			
	(T_{sat} = 223.99°C)				**(T_{sat} = 263.99°C)**			
20	1.0006	83.80	86.30	.2961	.9995	83.65	88.65	.2956
40	1.0067	167.25	169.77	.5715	1.0056	166.95	171.97	.5705
80	1.0280	334.29	336.86	1.0737	1.0268	333.72	338.85	1.0720
100	1.0423	418.24	420.85	1.3050	1.0410	417.52	422.72	1.3030
140	1.0784	587.82	590.52	1.7369	1.0768	586.76	592.15	1.7343
180	1.1261	761.16	763.97	2.1375	1.1240	759.63	765.25	2.1341
200	1.1555	849.9	852.8	2.3294	1.1530	848.1	853.9	2.3255
220	1.1898	940.7	943.7	2.5174	1.1866	938.4	944.4	2.5128
Sat.	1.1973	959.1	962.1	2.5546	1.2859	1147.8	1154.2	2.9202
	p = 75 bar = 7.5 MPa				**p = 100 bar = 10.0 MPa**			
	(T_{sat} = 290.59°C)				**(T_{sat} = 311.06°C)**			
20	.9984	83.50	90.99	.2950	.9972	83.36	93.33	.2945
40	1.0045	166.64	174.18	.5696	1.0034	166.35	176.38	.5686
80	1.0256	333.15	340.84	1.0704	1.0245	332.59	342.83	1.0688
100	1.0397	416.81	424.62	1.3011	1.0385	416.12	426.50	1.2992
140	1.0752	585.72	593.78	1.7317	1.0737	584.68	595.42	1.7292
180	1.1219	758.13	766.55	2.1308	1.1199	756.65	767.84	2.1275
220	1.1835	936.2	945.1	2.5083	1.1805	934.1	945.9	2.5039
260	1.2696	1124.4	1134.0	2.8763	1.2645	1121.1	1133.7	2.8699
Sat.	1.3677	1282.0	1292.2	3.1649	1.4524	1393.0	1407.6	3.3596
	p = 150 bar = 15.0 MPa				**p = 200 bar = 20.0 MPa**			
	(T_{sat} = 342.24°C)				**(T_{sat} = 365.81°C)**			
20	.9950	83.06	97.99	.2934	.9928	82.77	102.62	.2923
40	1.0013	165.76	180.78	.5666	.9992	165.17	185.16	.5646
80	1.0222	331.48	346.81	1.0656	1.0199	330.40	350.80	1.0624
100	1.0361	414.74	430.28	1.2955	1.0337	413.39	434.06	1.2917
140	1.0707	582.66	598.72	1.7242	1.0678	580.69	602.04	1.7193
180	1.1159	753.76	770.50	2.1210	1.1120	750.95	773.20	2.1147
220	1.1748	929.9	947.5	2.4953	1.1693	925.9	949.3	2.4870
260	1.2550	1114.6	1133.4	2.8576	1.2462	1108.6	1133.5	2.8459
300	1.3770	1316.6	1337.3	3.2260	1.3596	1306.1	1333.3	3.2071
Sat.	1.6581	1585.6	1610.5	3.6848	2.036	1785.6	1826.3	4.0139
	p = 250 bar = 25 MPa				**p = 300 bar = 30.0 MPa**			
20	.9907	82.47	107.24	.2911	.9886	82.17	111.84	.2899
40	.9971	164.60	189.52	.5626	.9951	164.04	193.89	.5607
100	1.0313	412.08	437.85	1.2881	1.0290	410.78	441.66	1.2844
200	1.1344	834.5	862.8	2.2961	1.1302	831.4	865.3	2.2893
300	1.3442	1296.6	1330.2	3.1900	1.3304	1287.9	1327.8	3.1741

v = (table value)/1000

v = (table value)/1000

TABLE A-6

Properties of Saturated Water (Solid–Vapor): Temperature Table

Pressure Conversions:
1 bar = 0.1 MPa
= 10² kPa

Temp. °C	Pressure kPa	Specific Volume m³/kg		Internal Energy kJ/kg			Enthalpy kJ/kg			Entropy kJ/kg · K		
		Sat. Solid $v_i \times 10^3$	Sat. Vapor v_g	Sat. Solid u_i	Subl. u_{ig}	Sat. Vapor u_g	Sat. Solid h_i	Subl. h_{ig}	Sat. Vapor h_g	Sat. Solid s_i	Subl. s_{ig}	Sat. Vapor s_g
.01	.6113	1.0908	206.1	−333.40	2708.7	2375.3	−333.40	2834.8	2501.4	−1.221	10.378	9.156
0	.6108	1.0908	206.3	−333.43	2708.8	2375.3	−333.43	2834.8	2501.3	−1.221	10.378	9.157
−2	.5176	1.0904	241.7	−337.62	2710.2	2372.6	−337.62	2835.3	2497.7	−1.237	10.456	9.219
−4	.4375	1.0901	283.8	−341.78	2711.6	2369.8	−341.78	2835.7	2494.0	−1.253	10.536	9.283
−6	.3689	1.0898	334.2	−345.91	2712.9	2367.0	−345.91	2836.2	2490.3	−1.268	10.616	9.348
−8	.3102	1.0894	394.4	−350.02	2714.2	2364.2	−350.02	2836.6	2486.6	−1.284	10.698	9.414
−10	.2602	1.0891	466.7	−354.09	2715.5	2361.4	−354.09	2837.0	2482.9	−1.299	10.781	9.481
−12	.2176	1.0888	553.7	−358.14	2716.8	2358.7	−358.14	2837.3	2479.2	−1.315	10.865	9.550
−14	.1815	1.0884	658.8	−362.15	2718.0	2355.9	−362.15	2837.6	2475.5	−1.331	10.950	9.619
−16	.1510	1.0881	786.0	−366.14	2719.2	2353.1	−366.14	2837.9	2471.8	−1.346	11.036	9.690
−18	.1252	1.0878	940.5	−370.10	2720.4	2350.3	−370.10	2838.2	2468.1	−1.362	11.123	9.762
−20	.1035	1.0874	1128.6	−374.03	2721.6	2347.5	−374.03	2838.4	2464.3	−1.377	11.212	9.835
−22	.0853	1.0871	1358.4	−377.93	2722.7	2344.7	−377.93	2838.6	2460.6	−1.393	11.302	9.909
−24	.0701	1.0868	1640.1	−381.80	2723.7	2342.0	−381.80	2838.7	2456.9	−1.408	11.394	9.985
−26	.0574	1.0864	1986.4	−385.64	2724.8	2339.2	−385.64	2838.9	2453.2	−1.424	11.486	10.062
−28	.0469	1.0861	2413.7	−389.45	2725.8	2336.4	−389.45	2839.0	2449.5	−1.439	11.580	10.141
−30	.0381	1.0858	2943	−393.23	2726.8	2333.6	−393.23	2839.0	2445.8	−1.455	11.676	10.221
−32	.0309	1.0854	3600	−396.98	2727.8	2330.8	−396.98	2839.1	2442.1	−1.471	11.773	10.303
−34	.0250	1.0851	4419	−400.71	2728.7	2328.0	−400.71	2839.1	2438.4	−1.486	11.872	10.386
−36	.0201	1.0848	5444	−404.40	2729.6	2325.2	−404.40	2839.1	2434.7	−1.501	11.972	10.470
−38	.0161	1.0844	6731	−408.06	2730.5	2322.4	−408.06	2839.0	2430.9	−1.517	12.073	10.556
−40	.0129	1.0841	8354	−411.70	2731.3	2319.6	−411.70	2838.9	2427.2	−1.532	12.176	10.644

$v = $ (table value)/1000

TABLE A-7

Pressure Conversions:
1 bar = 0.1 MPa
= 10² kPa

Properties of Saturated Refrigerant 22 (Liquid–Vapor): Temperature Table

R-22

Temp. °C	Press. bar	Specific Volume m^3/kg		Internal Energy kJ/kg		Enthalpy kJ/kg			Entropy $kJ/kg \cdot K$		Temp. °C
		Sat. Liquid $v_f \times 10^3$	Sat. Vapor v_g	Sat. Liquid u_f	Sat. Vapor u_g	Sat. Liquid h_f	Evap. h_{fg}	Sat. Vapor h_g	Sat. Liquid s_f	Sat. Vapor s_g	
−60	0.3749	0.6833	0.5370	−21.57	203.67	−21.55	245.35	223.81	−0.0964	1.0547	−60
−50	0.6451	0.6966	0.3239	−10.89	207.70	−10.85	239.44	228.60	−0.0474	1.0256	−50
−45	0.8290	0.7037	0.2564	−5.50	209.70	−5.44	236.39	230.95	−0.0235	1.0126	−45
−40	1.0522	0.7109	0.2052	−0.07	211.68	0.00	233.27	233.27	0.0000	1.0005	−40
−36	1.2627	0.7169	0.1730	4.29	213.25	4.38	230.71	235.09	0.0186	0.9914	−36
−32	1.5049	0.7231	0.1468	8.68	214.80	8.79	228.10	236.89	0.0369	0.9828	−32
−30	1.6389	0.7262	0.1355	10.88	215.58	11.00	226.77	237.78	0.0460	0.9787	−30
−28	1.7819	0.7294	0.1252	13.09	216.34	13.22	225.43	238.66	0.0551	0.9746	−28
−26	1.9345	0.7327	0.1159	15.31	217.11	15.45	224.08	239.53	0.0641	0.9707	−26
−22	2.2698	0.7393	0.0997	19.76	218.62	19.92	221.32	241.24	0.0819	0.9631	−22
−20	2.4534	0.7427	0.0926	21.99	219.37	22.17	219.91	242.09	0.0908	0.9595	−20
−18	2.6482	0.7462	0.0861	24.23	220.11	24.43	218.49	242.92	0.0996	0.9559	−18
−16	2.8547	0.7497	0.0802	26.48	220.85	26.69	217.05	243.74	0.1084	0.9525	−16
−14	3.0733	0.7533	0.0748	28.73	221.58	28.97	215.59	244.56	0.1171	0.9490	−14
−12	3.3044	0.7569	0.0698	31.00	222.30	31.25	214.11	245.36	0.1258	0.9457	−12
−10	3.5485	0.7606	0.0652	33.27	223.02	33.54	212.62	246.15	0.1345	0.9424	−10
−8	3.8062	0.7644	0.0610	35.54	223.73	35.83	211.10	246.93	0.1431	0.9392	−8
−6	4.0777	0.7683	0.0571	37.83	224.43	38.14	209.56	247.70	0.1517	0.9361	−6
−4	4.3638	0.7722	0.0535	40.12	225.13	40.46	208.00	248.45	0.1602	0.9330	−4
−2	4.6647	0.7762	0.0501	42.42	225.82	42.78	206.41	249.20	0.1688	0.9300	−2
0	4.9811	0.7803	0.0470	44.73	226.50	45.12	204.81	249.92	0.1773	0.9271	0
2	5.3133	0.7844	0.0442	47.04	227.17	47.46	203.18	250.64	0.1857	0.9241	2
4	5.6619	0.7887	0.0415	49.37	227.83	49.82	201.52	251.34	0.1941	0.9213	4
6	6.0275	0.7930	0.0391	51.71	228.48	52.18	199.84	252.03	0.2025	0.9184	6
8	6.4105	0.7974	0.0368	54.05	229.13	54.56	198.14	252.70	0.2109	0.9157	8
10	6.8113	0.8020	0.0346	56.40	229.76	56.95	196.40	253.35	0.2193	0.9129	10
12	7.2307	0.8066	0.0326	58.77	230.38	59.35	194.64	253.99	0.2276	0.9102	12
16	8.1268	0.8162	0.0291	63.53	231.59	64.19	191.02	255.21	0.2442	0.9048	16
20	9.1030	0.8263	0.0259	68.33	232.76	69.09	187.28	256.37	0.2607	0.8996	20
24	10.164	0.8369	0.0232	73.19	233.87	74.04	183.40	257.44	0.2772	0.8944	24
28	11.313	0.8480	0.0208	78.09	234.92	79.05	179.37	258.43	0.2936	0.8893	28
32	12.556	0.8599	0.0186	83.06	235.91	84.14	175.18	259.32	0.3101	0.8842	32
36	13.897	0.8724	0.0168	88.08	236.83	89.29	170.82	260.11	0.3265	0.8790	36
40	15.341	0.8858	0.0151	93.18	237.66	94.53	166.25	260.79	0.3429	0.8738	40
45	17.298	0.9039	0.0132	99.65	238.59	101.21	160.24	261.46	0.3635	0.8672	45
50	19.433	0.9238	0.0116	106.26	239.34	108.06	153.84	261.90	0.3842	0.8603	50
60	24.281	0.9705	0.0089	120.00	240.24	122.35	139.61	261.96	0.4264	0.8455	60

v_f = (table value)/1000

TABLE A-8

R-22

Pressure Conversions:
1 bar = 0.1 MPa
= 10² kPa

Properties of Saturated Refrigerant 22 (Liquid–Vapor): Pressure Table

Press. bar	Temp. °C	Specific Volume m³/kg Sat. Liquid $v_f \times 10^3$	Sat. Vapor v_g	Internal Energy kJ/kg Sat. Liquid u_f	Sat. Vapor u_g	Enthalpy kJ/kg Sat. Liquid h_f	Evap. h_{fg}	Sat. Vapor h_g	Entropy kJ/kg · K Sat. Liquid s_f	Sat. Vapor s_g	Press. bar
0.40	−58.86	0.6847	0.5056	−20.36	204.13	−20.34	244.69	224.36	−0.0907	1.0512	0.40
0.50	−54.83	0.6901	0.4107	−16.07	205.76	−16.03	242.33	226.30	−0.0709	1.0391	0.50
0.60	−51.40	0.6947	0.3466	−12.39	207.14	−12.35	240.28	227.93	−0.0542	1.0294	0.60
0.70	−48.40	0.6989	0.3002	−9.17	208.34	−9.12	238.47	229.35	−0.0397	1.0213	0.70
0.80	−45.73	0.7026	0.2650	−6.28	209.41	−6.23	236.84	230.61	−0.0270	1.0144	0.80
0.90	−43.30	0.7061	0.2374	−3.66	210.37	−3.60	235.34	231.74	−0.0155	1.0084	0.90
1.00	−41.09	0.7093	0.2152	−1.26	211.25	−1.19	233.95	232.77	−0.0051	1.0031	1.00
1.25	−36.23	0.7166	0.1746	4.04	213.16	4.13	230.86	234.99	0.0175	0.9919	1.25
1.50	−32.08	0.7230	0.1472	8.60	214.77	8.70	228.15	236.86	0.0366	0.9830	1.50
1.75	−28.44	0.7287	0.1274	12.61	216.18	12.74	225.73	238.47	0.0531	0.9755	1.75
2.00	−25.18	0.7340	0.1123	16.22	217.42	16.37	223.52	239.88	0.0678	0.9691	2.00
2.25	−22.22	0.7389	0.1005	19.51	218.53	19.67	221.47	241.15	0.0809	0.9636	2.25
2.50	−19.51	0.7436	0.0910	22.54	219.55	22.72	219.57	242.29	0.0930	0.9586	2.50
2.75	−17.00	0.7479	0.0831	25.36	220.48	25.56	217.77	243.33	0.1040	0.9542	2.75
3.00	−14.66	0.7521	0.0765	27.99	221.34	28.22	216.07	244.29	0.1143	0.9502	3.00
3.25	−12.46	0.7561	0.0709	30.47	222.13	30.72	214.46	245.18	0.1238	0.9465	3.25
3.50	−10.39	0.7599	0.0661	32.82	222.88	33.09	212.91	246.00	0.1328	0.9431	3.50
3.75	−8.43	0.7636	0.0618	35.06	223.58	35.34	211.42	246.77	0.1413	0.9399	3.75
4.00	−6.56	0.7672	0.0581	37.18	224.24	37.49	209.99	247.48	0.1493	0.9370	4.00
4.25	−4.78	0.7706	0.0548	39.22	224.86	39.55	208.61	248.16	0.1569	0.9342	4.25
4.50	−3.08	0.7740	0.0519	41.17	225.45	41.52	207.27	248.80	0.1642	0.9316	4.50
4.75	−1.45	0.7773	0.0492	43.05	226.00	43.42	205.98	249.40	0.1711	0.9292	4.75
5.00	0.12	0.7805	0.0469	44.86	226.54	45.25	204.71	249.97	0.1777	0.9269	5.00
5.25	1.63	0.7836	0.0447	46.61	227.04	47.02	203.48	250.51	0.1841	0.9247	5.25
5.50	3.08	0.7867	0.0427	48.30	227.53	48.74	202.28	251.02	0.1903	0.9226	5.50
5.75	4.49	0.7897	0.0409	49.94	227.99	50.40	201.11	251.51	0.1962	0.9206	5.75
6.00	5.85	0.7927	0.0392	51.53	228.44	52.01	199.97	251.98	0.2019	0.9186	6.00
7.00	10.91	0.8041	0.0337	57.48	230.04	58.04	195.60	253.64	0.2231	0.9117	7.00
8.00	15.45	0.8149	0.0295	62.88	231.43	63.53	191.52	255.05	0.2419	0.9056	8.00
9.00	19.59	0.8252	0.0262	67.84	232.64	68.59	187.67	256.25	0.2591	0.9001	9.00
10.00	23.40	0.8352	0.0236	72.46	233.71	73.30	183.99	257.28	0.2748	0.8952	10.00
12.00	30.25	0.8546	0.0195	80.87	235.48	81.90	177.04	258.94	0.3029	0.8864	12.00
14.00	36.29	0.8734	0.0166	88.45	236.89	89.68	170.49	260.16	0.3277	0.8786	14.00
16.00	41.73	0.8919	0.0144	95.41	238.00	96.83	164.21	261.04	0.3500	0.8715	16.00
18.00	46.69	0.9104	0.0127	101.87	238.86	103.51	158.13	261.64	0.3705	0.8649	18.00
20.00	51.26	0.9291	0.0112	107.95	239.51	109.81	152.17	261.98	0.3895	0.8586	20.00
24.00	59.46	0.9677	0.0091	119.24	240.22	121.56	140.43	261.99	0.4241	0.8463	24.00

v_f = (table value)/1000

TABLE A-9

Properties of Superheated Refrigerant 22 Vapor

R-22

T °C	v m³/kg	u kJ/kg	h kJ/kg	s kJ/kg·K	v m³/kg	u kJ/kg	h kJ/kg	s kJ/kg·K
	p = 0.4 bar = 0.04 MPa (T_{sat} = −58.86°C)				p = 0.6 bar = 0.06 MPa (T_{sat} = −51.40°C)			
Sat.	0.50559	204.13	224.36	1.0512	0.34656	207.14	227.93	1.0294
−55	0.51532	205.92	226.53	1.0612				
−50	0.52787	208.26	229.38	1.0741	0.34895	207.80	228.74	1.0330
−45	0.54037	210.63	232.24	1.0868	0.35747	210.20	231.65	1.0459
−40	0.55284	213.02	235.13	1.0993	0.36594	212.62	234.58	1.0586
−35	0.56526	215.43	238.05	1.1117	0.37437	215.06	237.52	1.0711
−30	0.57766	217.88	240.99	1.1239	0.38277	217.53	240.49	1.0835
−25	0.59002	220.35	243.95	1.1360	0.39114	220.02	243.49	1.0956
−20	0.60236	222.85	246.95	1.1479	0.39948	222.54	246.51	1.1077
−15	0.61468	225.38	249.97	1.1597	0.40779	225.08	249.55	1.1196
−10	0.62697	227.93	253.01	1.1714	0.41608	227.65	252.62	1.1314
−5	0.63925	230.52	256.09	1.1830	0.42436	230.25	255.71	1.1430
0	0.65151	233.13	259.19	1.1944	0.43261	232.88	258.83	1.1545
	p = 0.8 bar = 0.08 MPa (T_{sat} = −45.73°C)				p = 1.0 bar = 0.10 MPa (T_{sat} = −41.09°C)			
Sat.	0.26503	209.41	230.61	1.0144	0.21518	211.25	232.77	1.0031
−45	0.26597	209.76	231.04	1.0163				
−40	0.27245	212.21	234.01	1.0292	0.21633	211.79	233.42	1.0059
−35	0.27890	214.68	236.99	1.0418	0.22158	214.29	236.44	1.0187
−30	0.28530	217.17	239.99	1.0543	0.22679	216.80	239.48	1.0313
−25	0.29167	219.68	243.02	1.0666	0.23197	219.34	242.54	1.0438
−20	0.29801	222.22	246.06	1.0788	0.23712	221.90	245.61	1.0560
−15	0.30433	224.78	249.13	1.0908	0.24224	224.48	248.70	1.0681
−10	0.31062	227.37	252.22	1.1026	0.24734	227.08	251.82	1.0801
−5	0.31690	229.98	255.34	1.1143	0.25241	229.71	254.95	1.0919
0	0.32315	232.62	258.47	1.1259	0.25747	232.36	258.11	1.1035
5	0.32939	235.29	261.64	1.1374	0.26251	235.04	261.29	1.1151
10	0.33561	237.98	264.83	1.1488	0.26753	237.74	264.50	1.1265
	p = 1.5 bar = 0.15 MPa (T_{sat} = −32.08°C)				p = 2.0 bar = 0.20 MPa (T_{sat} = −25.18°C)			
Sat.	0.14721	214.77	236.86	0.9830	0.11232	217.42	239.88	0.9691
−30	0.14872	215.85	238.16	0.9883				
−25	0.15232	218.45	241.30	1.0011	0.11242	217.51	240.00	0.9696
−20	0.15588	221.07	244.45	1.0137	0.11520	220.19	243.23	0.9825
−15	0.15941	223.70	247.61	1.0260	0.11795	222.88	246.47	0.9952
−10	0.16292	226.35	250.78	1.0382	0.12067	225.58	249.72	1.0076
−5	0.16640	229.02	253.98	1.0502	0.12336	228.30	252.97	1.0199
0	0.16987	231.70	257.18	1.0621	0.12603	231.03	256.23	1.0310
5	0.17331	234.42	260.41	1.0738	0.12868	233.78	259.51	1.0438
10	0.17674	237.15	263.66	1.0854	0.13132	236.54	262.81	1.0555
15	0.18015	239.91	266.93	1.0968	0.13393	239.33	266.12	1.0671
20	0.18355	242.69	270.22	1.1081	0.13653	242.14	269.44	1.0786
25	0.18693	245.49	273.53	1.1193	0.13912	244.97	272.79	1.0899

TABLE A-9

(Continued)

T °C	v m³/kg	u kJ/kg	h kJ/kg	s kJ/kg·K	v m³/kg	u kJ/kg	h kJ/kg	s kJ/kg·K
	p = 2.5 bar = 0.25 MPa (T_sat = −19.51°C)				p = 3.0 bar = 0.30 MPa (T_sat = −14.66°C)			
Sat.	0.09097	219.55	242.29	0.9586	0.07651	221.34	244.29	0.9502
−15	0.09303	222.03	245.29	0.9703				
−10	0.09528	224.79	248.61	0.9831	0.07833	223.96	247.46	0.9623
−5	0.09751	227.55	251.93	0.9956	0.08025	226.78	250.86	0.9751
0	0.09971	230.33	255.26	1.0078	0.08214	229.61	254.25	0.9876
5	0.10189	233.12	258.59	1.0199	0.08400	232.44	257.64	0.9999
10	0.10405	235.92	261.93	1.0318	0.08585	235.28	261.04	1.0120
15	0.10619	238.74	265.29	1.0436	0.08767	238.14	264.44	1.0239
20	0.10831	241.58	268.66	1.0552	0.08949	241.01	267.85	1.0357
25	0.11043	244.44	272.04	1.0666	0.09128	243.89	271.28	1.0472
30	0.11253	247.31	275.44	1.0779	0.09307	246.80	274.72	1.0587
35	0.11461	250.21	278.86	1.0891	0.09484	249.72	278.17	1.0700
40	0.11669	253.13	282.30	1.1002	0.09660	252.66	281.64	1.0811
	p = 3.5 bar = 0.35 MPa (T_sat = −10.39°C)				p = 4.0 bar = 0.40 MPa (T_sat = −6.56°C)			
Sat.	0.06605	222.88	246.00	0.9431	0.05812	224.24	247.48	0.9370
−10	0.06619	223.10	246.27	0.9441				
−5	0.06789	225.99	249.75	0.9572	0.05860	225.16	248.60	0.9411
0	0.06956	228.86	253.21	0.9700	0.06011	228.09	252.14	0.9542
5	0.07121	231.74	256.67	0.9825	0.06160	231.02	225.66	0.9670
10	0.07284	234.63	260.12	0.9948	0.06306	233.95	259.18	0.9795
15	0.07444	237.52	263.57	1.0069	0.06450	236.89	262.69	0.9918
20	0.07603	240.42	267.03	1.0188	0.06592	239.83	266.19	1.0039
25	0.07760	243.34	270.50	1.0305	0.06733	242.77	269.71	1.0158
30	0.07916	246.27	273.97	1.0421	0.06872	245.73	273.22	1.0274
35	0.08070	249.22	227.46	1.0535	0.07010	248.71	276.75	1.0390
40	0.08224	252.18	280.97	1.0648	0.07146	251.70	280.28	1.0504
45	0.08376	255.17	284.48	1.0759	0.07282	254.70	283.83	1.0616
	p = 4.5 bar = 0.45 MPa (T_sat = −3.08°C)				p = 5.0 bar = 0.50 MPa (T_sat = 0.12°C)			
Sat.	0.05189	225.45	248.80	0.9316	0.04686	226.54	249.97	0.9269
0	0.05275	227.29	251.03	0.9399				
5	0.05411	230.28	254.63	0.9529	0.04810	229.52	253.57	0.9399
10	0.05545	233.26	258.21	0.9657	0.04934	232.55	257.22	0.9530
15	0.05676	236.24	261.78	0.9782	0.05056	235.57	260.85	0.9657
20	0.05805	239.22	265.34	0.9904	0.05175	238.59	264.47	0.9781
25	0.05933	242.20	268.90	1.0025	0.05293	241.61	268.07	0.9903
30	0.06059	245.19	272.46	1.0143	0.05409	244.63	271.68	1.0023
35	0.06184	248.19	276.02	1.0259	0.05523	247.66	275.28	1.0141
40	0.06308	251.20	279.59	1.0374	0.05636	250.70	278.89	1.0257
45	0.06430	254.23	283.17	1.0488	0.05748	253.76	282.50	1.0371
50	0.06552	257.28	286.76	1.0600	0.05859	256.82	286.12	1.0484
55	0.06672	260.34	290.36	1.0710	0.05969	259.90	289.75	1.0595

TABLE A-9

(Continued)

T °C	v m³/kg	u kJ/kg	h kJ/kg	s kJ/kg·K	v m³/kg	u kJ/kg	h kJ/kg	s kJ/kg·K
	p = 5.5 bar = 0.55 MPa (T_{sat} = 3.08°C)				p = 6.0 bar = 0.60 MPa (T_{sat} = 5.85°C)			
Sat.	0.04271	227.53	251.02	0.9226	0.03923	228.44	251.98	0.9186
5	0.04317	228.72	252.46	0.9278				
10	0.04433	231.81	256.20	0.9411	0.04015	231.05	255.14	0.9299
15	0.04547	234.89	259.90	0.9540	0.04122	234.18	258.91	0.9431
20	0.04658	237.95	263.57	0.9667	0.04227	237.29	262.65	0.9560
25	0.04768	241.01	267.23	0.9790	0.04330	240.39	266.37	0.9685
30	0.04875	244.07	270.88	0.9912	0.04431	243.49	270.07	0.9808
35	0.04982	247.13	274.53	1.0031	0.04530	246.58	273.76	0.9929
40	0.05086	250.20	278.17	1.0148	0.04628	249.68	277.45	1.0048
45	0.05190	253.27	281.82	1.0264	0.04724	252.78	281.13	1.0164
50	0.05293	256.36	285.47	1.0378	0.04820	255.90	284.82	1.0279
55	0.05394	259.46	289.13	1.0490	0.04914	259.02	288.51	1.0393
60	0.05495	262.58	292.80	1.0601	0.05008	262.15	292.20	1.0504
	p = 7.0 bar = 0.70 MPa (T_{sat} = 10.91°C)				p = 8.0 bar = 0.80 MPa (T_{sat} = 15.45°C)			
Sat.	0.03371	230.04	253.64	0.9117	0.02953	231.43	255.05	0.9056
15	0.03451	232.70	256.86	0.9229				
20	0.03547	235.92	260.75	0.9363	0.03033	234.47	258.74	0.9182
25	0.03639	239.12	264.59	0.9493	0.03118	237.76	262.70	0.9315
30	0.03730	242.29	268.40	0.9619	0.03202	241.04	266.66	0.9448
35	0.03819	245.46	272.19	0.9743	0.03283	244.28	270.54	0.9574
40	0.03906	248.62	275.96	0.9865	0.03363	247.52	274.42	0.9700
45	0.03992	251.78	279.72	0.9984	0.03440	250.74	278.26	0.9821
50	0.04076	254.94	283.48	1.0101	0.03517	253.96	282.10	0.9941
55	0.04160	258.11	287.23	1.0216	0.03592	257.18	285.92	1.0058
60	0.04242	261.29	290.99	1.0330	0.03667	260.40	289.74	1.0174
65	0.04324	264.48	294.75	1.0442	0.03741	263.64	293.56	1.0287
70	0.04405	267.68	298.51	1.0552	0.03814	266.87	297.38	1.0400
	p = 9.0 bar = 0.90 MPa (T_{sat} = 19.59°C)				p = 10.0 bar = 1.00 MPa (T_{sat} = 23.40°C)			
Sat.	0.02623	232.64	256.25	0.9001	0.02358	233.71	257.28	0.8952
20	0.02630	232.92	256.59	0.9013				
30	0.02789	239.73	264.83	0.9289	0.02457	238.34	262.91	0.9139
40	0.02939	246.37	272.82	0.9549	0.02598	245.18	271.17	0.9407
50	0.03082	252.95	280.68	0.9795	0.02732	251.90	279.22	0.9660
60	0.03219	259.49	288.46	1.0033	0.02860	258.56	287.15	0.9902
70	0.03353	266.04	296.21	1.0262	0.02984	265.19	295.03	1.0135
80	0.03483	272.62	303.96	1.0484	0.03104	271.84	302.88	1.0361
90	0.03611	279.23	311.73	1.0701	0.03221	278.52	310.74	1.0580
100	0.03736	285.90	319.53	1.0913	0.03337	285.24	318.61	1.0794
110	0.03860	292.63	327.37	1.1120	0.03450	292.02	326.52	1.1003
120	0.03982	299.42	335.26	1.1323	0.03562	298.85	334.46	1.1207
130	0.04103	306.28	343.21	1.1523	0.03672	305.74	342.46	1.1408
140	0.04223	313.21	351.22	1.1719	0.03781	312.70	350.51	1.1605
150	0.04342	320.21	359.29	1.1912	0.03889	319.74	358.63	1.1790

(Continued)

T °C	v m³/kg	u kJ/kg	h kJ/kg	s kJ/kg·K	v m³/kg	u kJ/kg	h kJ/kg	s kJ/kg·K
	p = 12.0 bar = 1.20 MPa (T_sat = 30.25°C)				p = 14.0 bar = 1.40 MPa (T_sat = 36.29°C)			
Sat.	0.01955	235.48	258.94	0.8864	0.01662	236.89	260.16	0.8786
40	0.02083	242.63	267.62	0.9146	0.01708	239.78	263.70	0.8900
50	0.02204	249.69	276.14	0.9413	0.01823	247.29	272.81	0.9186
60	0.02319	256.60	284.43	0.9666	0.01929	254.52	281.53	0.9452
70	0.02428	263.44	292.58	0.9907	0.02029	261.60	290.01	0.9703
80	0.02534	270.25	300.66	1.0139	0.02125	268.60	298.34	0.9942
90	0.02636	277.07	308.70	1.0363	0.02217	275.56	306.60	1.0172
100	0.02736	283.90	316.73	1.0582	0.02306	282.52	314.80	1.0395
110	0.02834	290.77	324.78	1.0794	0.02393	289.49	323.00	1.0612
120	0.02930	297.69	332.85	1.1002	0.02478	296.50	331.19	1.0823
130	0.03024	304.65	340.95	1.1205	0.02562	303.55	339.41	1.1029
140	0.03118	311.68	349.09	1.1405	0.02644	310.64	347.65	1.1231
150	0.03210	318.77	357.29	1.1601	0.02725	317.79	355.94	1.1429
160	0.03301	325.92	365.54	1.1793	0.02805	324.99	364.26	1.1624
170	0.03392	333.14	373.84	1.1983	0.02884	332.26	372.64	1.1815

T °C	v m³/kg	u kJ/kg	h kJ/kg	s kJ/kg·K	v m³/kg	u kJ/kg	h kJ/kg	s kJ/kg·K
	p = 16.0 bar = 1.60 MPa (T_sat = 41.73°C)				p = 18.0 bar = 1.80 MPa (T_sat = 46.69°C)			
Sat.	0.01440	238.00	261.04	0.8715	0.01265	238.86	261.64	0.8649
50	0.01533	244.66	269.18	0.8971	0.01301	241.72	265.14	0.8758
60	0.01634	252.29	278.43	0.9252	0.01401	249.86	275.09	0.9061
70	0.01728	259.65	287.30	0.9515	0.01492	257.57	284.43	0.9337
80	0.01817	266.86	295.93	0.9762	0.01576	265.04	293.40	0.9595
90	0.01901	274.00	304.42	0.9999	0.01655	272.37	302.16	0.9839
100	0.01983	281.09	312.82	1.0228	0.01731	279.62	310.77	1.0073
110	0.02062	288.18	321.17	1.0448	0.01804	286.83	319.30	1.0299
120	0.02139	295.28	329.51	1.0663	0.01874	294.04	327.78	1.0517
130	0.02214	302.41	337.84	1.0872	0.01943	301.26	336.24	1.0730
140	0.02288	309.58	346.19	1.1077	0.02011	308.50	344.70	1.0937
150	0.02361	316.79	354.56	1.1277	0.02077	315.78	353.17	1.1139
160	0.02432	324.05	362.97	1.1473	0.02142	323.10	361.66	1.1338
170	0.02503	331.37	371.42	1.1666	0.02207	330.47	370.19	1.1532

T °C	v m³/kg	u kJ/kg	h kJ/kg	s kJ/kg·K	v m³/kg	u kJ/kg	h kJ/kg	s kJ/kg·K
	p = 20.0 bar = 2.00 MPa (T_sat = 51.26°C)				p = 24.0 bar = 2.4 MPa (T_sat = 59.46°C)			
Sat.	0.01124	239.51	261.98	0.8586	0.00907	240.22	261.99	0.8463
60	0.01212	247.20	271.43	0.8873	0.00913	240.78	262.68	0.8484
70	0.01300	255.35	281.36	0.9167	0.01006	250.30	274.43	0.8831
80	0.01381	263.12	290.74	0.9436	0.01085	258.89	284.93	0.9133
90	0.01457	270.67	299.80	0.9689	0.01156	267.01	294.75	0.9407
100	0.01528	278.09	308.65	0.9929	0.01222	274.85	304.18	0.9663
110	0.01596	285.44	317.37	1.0160	0.01284	282.53	313.35	0.9906
120	0.01663	292.76	326.01	1.0383	0.01343	290.11	322.35	1.0137
130	0.01727	300.08	334.61	1.0598	0.01400	297.64	331.25	1.0361
140	0.01789	307.40	343.19	1.0808	0.01456	305.14	340.08	1.0577
150	0.01850	314.75	351.76	1.1013	0.01509	312.64	348.87	1.0787
160	0.01910	322.14	360.34	1.1214	0.01562	320.16	357.64	1.0992
170	0.01969	329.56	368.95	1.1410	0.01613	327.70	366.41	1.1192
180	0.02027	337.03	377.58	1.1603	0.01663	335.27	375.20	1.1388

TABLE A-10

Properties of Saturated Refrigerant 134a (Liquid–Vapor): Temperature Table

Pressure Conversions:
1 bar = 0.1 MPa
= 10^2 kPa

Temp. °C	Press. bar	Specific Volume m^3/kg Sat. Liquid $v_f \times 10^3$	Sat. Vapor v_g	Internal Energy kJ/kg Sat. Liquid u_f	Sat. Vapor u_g	Enthalpy kJ/kg Sat. Liquid h_f	Evap. h_{fg}	Sat. Vapor h_g	Entropy kJ/kg·K Sat. Liquid s_f	Sat. Vapor s_g	Temp. °C
−40	0.5164	0.7055	0.3569	−0.04	204.45	0.00	222.88	222.88	0.0000	0.9560	−40
−36	0.6332	0.7113	0.2947	4.68	206.73	4.73	220.67	225.40	0.0201	0.9506	−36
−32	0.7704	0.7172	0.2451	9.47	209.01	9.52	218.37	227.90	0.0401	0.9456	−32
−28	0.9305	0.7233	0.2052	14.31	211.29	14.37	216.01	230.38	0.0600	0.9411	−28
−26	1.0199	0.7265	0.1882	16.75	212.43	16.82	214.80	231.62	0.0699	0.9390	−26
−24	1.1160	0.7296	0.1728	19.21	213.57	19.29	213.57	232.85	0.0798	0.9370	−24
−22	1.2192	0.7328	0.1590	21.68	214.70	21.77	212.32	234.08	0.0897	0.9351	−22
−20	1.3299	0.7361	0.1464	24.17	215.84	24.26	211.05	235.31	0.0996	0.9332	−20
−18	1.4483	0.7395	0.1350	26.67	216.97	26.77	209.76	236.53	0.1094	0.9315	−18
−16	1.5748	0.7428	0.1247	29.18	218.10	29.30	208.45	237.74	0.1192	0.9298	−16
−12	1.8540	0.7498	0.1068	34.25	220.36	34.39	205.77	240.15	0.1388	0.9267	−12
−8	2.1704	0.7569	0.0919	39.38	222.60	39.54	203.00	242.54	0.1583	0.9239	−8
−4	2.5274	0.7644	0.0794	44.56	224.84	44.75	200.15	244.90	0.1777	0.9213	−4
0	2.9282	0.7721	0.0689	49.79	227.06	50.02	197.21	247.23	0.1970	0.9190	0
4	3.3765	0.7801	0.0600	55.08	229.27	55.35	194.19	249.53	0.2162	0.9169	4
8	3.8756	0.7884	0.0525	60.43	231.46	60.73	191.07	251.80	0.2354	0.9150	8
12	4.4294	0.7971	0.0460	65.83	233.63	66.18	187.85	254.03	0.2545	0.9132	12
16	5.0416	0.8062	0.0405	71.29	235.78	71.69	184.52	256.22	0.2735	0.9116	16
20	5.7160	0.8157	0.0358	76.80	237.91	77.26	181.09	258.36	0.2924	0.9102	20
24	6.4566	0.8257	0.0317	82.37	240.01	82.90	177.55	260.45	0.3113	0.9089	24
26	6.8530	0.8309	0.0298	85.18	241.05	85.75	175.73	261.48	0.3208	0.9082	26
28	7.2675	0.8362	0.0281	88.00	242.08	88.61	173.89	262.50	0.3302	0.9076	28
30	7.7006	0.8417	0.0265	90.84	243.10	91.49	172.00	263.50	0.3396	0.9070	30
32	8.1528	0.8473	0.0250	93.70	244.12	94.39	170.09	264.48	0.3490	0.9064	32
34	8.6247	0.8530	0.0236	96.58	245.12	97.31	168.14	265.45	0.3584	0.9058	34
36	9.1168	0.8590	0.0223	99.47	246.11	100.25	166.15	266.40	0.3678	0.9053	36
38	9.6298	0.8651	0.0210	102.38	247.09	103.21	164.12	267.33	0.3772	0.9047	38
40	10.164	0.8714	0.0199	105.30	248.06	106.19	162.05	268.24	0.3866	0.9041	40
42	10.720	0.8780	0.0188	108.25	249.02	109.19	159.94	269.14	0.3960	0.9035	42
44	11.299	0.8847	0.0177	111.22	249.96	112.22	157.79	270.01	0.4054	0.9030	44
48	12.526	0.8989	0.0159	117.22	251.79	118.35	153.33	271.68	0.4243	0.9017	48
52	13.851	0.9142	0.0142	123.31	253.55	124.58	148.66	273.24	0.4432	0.9004	52
56	15.278	0.9308	0.0127	129.51	255.23	130.93	143.75	274.68	0.4622	0.8990	56
60	16.813	0.9488	0.0114	135.82	256.81	137.42	138.57	275.99	0.4814	0.8973	60
70	21.162	1.0027	0.0086	152.22	260.15	154.34	124.08	278.43	0.5302	0.8918	70
80	26.324	1.0766	0.0064	169.88	262.14	172.71	106.41	279.12	0.5814	0.8827	80
90	32.435	1.1949	0.0046	189.82	261.34	193.69	82.63	276.32	0.6380	0.8655	90
100	39.742	1.5443	0.0027	218.60	248.49	224.74	34.40	259.13	0.7196	0.8117	100

R-134a

v_f = (table value)/1000

TABLE A-11

R-134a

Properties of Saturated Refrigerant 134a (Liquid–Vapor): Pressure Table

Pressure Conversions:
1 bar = 0.1 MPa
= 10^2 kPa

Press. bar	Temp. °C	Specific Volume m^3/kg		Internal Energy kJ/kg		Enthalpy kJ/kg			Entropy kJ/kg · K		Press. bar
		Sat. Liquid $v_f \times 10^3$	Sat. Vapor v_g	Sat. Liquid u_f	Sat. Vapor u_g	Sat. Liquid h_f	Evap. h_{fg}	Sat. Vapor h_g	Sat. Liquid s_f	Sat. Vapor s_g	
0.6	−37.07	0.7097	0.3100	3.41	206.12	3.46	221.27	224.72	0.0147	0.9520	0.6
0.8	−31.21	0.7184	0.2366	10.41	209.46	10.47	217.92	228.39	0.0440	0.9447	0.8
1.0	−26.43	0.7258	0.1917	16.22	212.18	16.29	215.06	231.35	0.0678	0.9395	1.0
1.2	−22.36	0.7323	0.1614	21.23	214.50	21.32	212.54	233.86	0.0879	0.9354	1.2
1.4	−18.80	0.7381	0.1395	25.66	216.52	25.77	210.27	236.04	0.1055	0.9322	1.4
1.6	−15.62	0.7435	0.1229	29.66	218.32	29.78	208.19	237.97	0.1211	0.9295	1.6
1.8	−12.73	0.7485	0.1098	33.31	219.94	33.45	206.26	239.71	0.1352	0.9273	1.8
2.0	−10.09	0.7532	0.0993	36.69	221.43	36.84	204.46	241.30	0.1481	0.9253	2.0
2.4	−5.37	0.7618	0.0834	42.77	224.07	42.95	201.14	244.09	0.1710	0.9222	2.4
2.8	−1.23	0.7697	0.0719	48.18	226.38	48.39	198.13	246.52	0.1911	0.9197	2.8
3.2	2.48	0.7770	0.0632	53.06	228.43	53.31	195.35	248.66	0.2089	0.9177	3.2
3.6	5.84	0.7839	0.0564	57.54	230.28	57.82	192.76	250.58	0.2251	0.9160	3.6
4.0	8.93	0.7904	0.0509	61.69	231.97	62.00	190.32	252.32	0.2399	0.9145	4.0
5.0	15.74	0.8056	0.0409	70.93	235.64	71.33	184.74	256.07	0.2723	0.9117	5.0
6.0	21.58	0.8196	0.0341	78.99	238.74	79.48	179.71	259.19	0.2999	0.9097	6.0
7.0	26.72	0.8328	0.0292	86.19	241.42	86.78	175.07	261.85	0.3242	0.9080	7.0
8.0	31.33	0.8454	0.0255	92.75	243.78	93.42	170.73	264.15	0.3459	0.9066	8.0
9.0	35.53	0.8576	0.0226	98.79	245.88	99.56	166.62	266.18	0.3656	0.9054	9.0
10.0	39.39	0.8695	0.0202	104.42	247.77	105.29	162.68	267.97	0.3838	0.9043	10.0
12.0	46.32	0.8928	0.0166	114.69	251.03	115.76	155.23	270.99	0.4164	0.9023	12.0
14.0	52.43	0.9159	0.0140	123.98	253.74	125.26	148.14	273.40	0.4453	0.9003	14.0
16.0	57.92	0.9392	0.0121	132.52	256.00	134.02	141.31	275.33	0.4714	0.8982	16.0
18.0	62.91	0.9631	0.0105	140.49	257.88	142.22	134.60	276.83	0.4954	0.8959	18.0
20.0	67.49	0.9878	0.0093	148.02	259.41	149.99	127.95	277.94	0.5178	0.8934	20.0
25.0	77.59	1.0562	0.0069	165.48	261.84	168.12	111.06	279.17	0.5687	0.8854	25.0
30.0	86.22	1.1416	0.0053	181.88	262.16	185.30	92.71	278.01	0.6156	0.8735	30.0

v_f = (table value)/1000

Properties of Superheated Refrigerant 134a Vapor

T °C	v m³/kg	u kJ/kg	h kJ/kg	s kJ/kg·K	v m³/kg	u kJ/kg	h kJ/kg	s kJ/kg·K
	$p = 0.6$ bar = 0.06 MPa ($T_{sat} = -37.07°C$)				$p = 1.0$ bar = 0.10 MPa ($T_{sat} = -26.43°C$)			
Sat.	0.31003	206.12	224.72	0.9520	0.19170	212.18	231.35	0.9395
−20	0.33536	217.86	237.98	1.0062	0.19770	216.77	236.54	0.9602
−10	0.34992	224.97	245.96	1.0371	0.20686	224.01	244.70	0.9918
0	0.36433	232.24	254.10	1.0675	0.21587	231.41	252.99	1.0227
10	0.37861	239.69	262.41	1.0973	0.22473	238.96	261.43	1.0531
20	0.39279	247.32	270.89	1.1267	0.23349	246.67	270.02	1.0829
30	0.40688	255.12	279.53	1.1557	0.24216	254.54	278.76	1.1122
40	0.42091	263.10	288.35	1.1844	0.25076	262.58	287.66	1.1411
50	0.43487	271.25	297.34	1.2126	0.25930	270.79	296.72	1.1696
60	0.44879	279.58	306.51	1.2405	0.26779	279.16	305.94	1.1977
70	0.46266	288.08	315.84	1.2681	0.27623	287.70	315.32	1.2254
80	0.47650	296.75	325.34	1.2954	0.28464	296.40	324.87	1.2528
90	0.49031	305.58	335.00	1.3224	0.29302	305.27	334.57	1.2799
	$p = 1.4$ bar = 0.14 MPa ($T_{sat} = -18.80°C$)				$p = 1.8$ bar = 0.18 MPa ($T_{sat} = -12.73°C$)			
Sat.	0.13945	216.52	236.04	0.9322	0.10983	219.94	239.71	0.9273
−10	0.14549	223.03	243.40	0.9606	0.11135	222.02	242.06	0.9362
0	0.15219	230.55	251.86	0.9922	0.11678	229.67	250.69	0.9684
10	0.15875	238.21	260.43	1.0230	0.12207	237.44	259.41	0.9998
20	0.16520	246.01	269.13	1.0532	0.12723	245.33	268.23	1.0304
30	0.17155	253.96	277.97	1.0828	0.13230	253.36	277.17	1.0604
40	0.17783	262.06	286.96	1.1120	0.13730	261.53	286.24	1.0898
50	0.18404	270.32	296.09	1.1407	0.14222	269.85	295.45	1.1187
60	0.19020	278.74	305.37	1.1690	0.14710	278.31	304.79	1.1472
70	0.19633	287.32	314.80	1.1969	0.15193	286.93	314.28	1.1753
80	0.20241	296.06	324.39	1.2244	0.15672	295.71	323.92	1.2030
90	0.20846	304.95	334.14	1.2516	0.16148	304.63	333.70	1.2303
100	0.21449	314.01	344.04	1.2785	0.16622	313.72	343.63	1.2573
	$p = 2.0$ bar = 0.20 MPa ($T_{sat} = -10.09°C$)				$p = 2.4$ bar = 0.24 MPa ($T_{sat} = -5.37°C$)			
Sat.	0.09933	221.43	241.30	0.9253	0.08343	224.07	244.09	0.9222
−10	0.09938	221.50	241.38	0.9256				
0	0.10438	229.23	250.10	0.9582	0.08574	228.31	248.89	0.9399
10	0.10922	237.05	258.89	0.9898	0.08993	236.26	257.84	0.9721
20	0.11394	244.99	267.78	1.0206	0.09399	244.30	266.85	1.0034
30	0.11856	253.06	276.77	1.0508	0.09794	252.45	275.95	1.0339
40	0.12311	261.26	285.88	1.0804	0.10181	260.72	285.16	1.0637
50	0.12758	269.61	295.12	1.1094	0.10562	269.12	294.47	1.0930
60	0.13201	278.10	304.50	1.1380	0.10937	277.67	303.91	1.1218
70	0.13639	286.74	314.02	1.1661	0.11307	286.35	313.49	1.1501
80	0.14073	295.53	323.68	1.1939	0.11674	295.18	323.19	1.1780
90	0.14504	304.47	333.48	1.2212	0.12037	304.15	333.04	1.2055
100	0.14932	313.57	343.43	1.2483	0.12398	313.27	343.03	1.2326

Pressure Conversions:
1 bar = 0.1 MPa
= 10² kPa

R-134a

(Continued)

Pressure Conversions:
1 bar = 0.1 MPa
= 10^2 kPa

T °C	v m³/kg	u kJ/kg	h kJ/kg	s kJ/kg·K	v m³/kg	u kJ/kg	h kJ/kg	s kJ/kg·K
	p = 2.8 bar = 0.28 MPa (T_{sat} = −1.23°C)				p = 3.2 bar = 0.32 MPa (T_{sat} = 2.48°C)			
Sat.	0.07193	226.38	246.52	0.9197	0.06322	228.43	248.66	0.9177
0	0.07240	227.37	247.64	0.9238				
10	0.07613	235.44	256.76	0.9566	0.06576	234.61	255.65	0.9427
20	0.07972	243.59	265.91	0.9883	0.06901	242.87	264.95	0.9749
30	0.08320	251.83	275.12	1.0192	0.07214	251.19	274.28	1.0062
40	0.08660	260.17	284.42	1.0494	0.07518	259.61	283.67	1.0367
50	0.08992	268.64	293.81	1.0789	0.07815	268.14	293.15	1.0665
60	0.09319	277.23	303.32	1.1079	0.08106	276.79	302.72	1.0957
70	0.09641	285.96	312.95	1.1364	0.08392	285.56	312.41	1.1243
80	0.09960	294.82	322.71	1.1644	0.08674	294.46	322.22	1.1525
90	0.10275	303.83	332.60	1.1920	0.08953	303.50	332.15	1.1802
100	0.10587	312.98	342.62	1.2193	0.09229	312.68	342.21	1.2076
110	0.10897	322.27	352.78	1.2461	0.09503	322.00	352.40	1.2345
120	0.11205	331.71	363.08	1.2727	0.09774	331.45	362.73	1.2611
	p = 4.0 bar = 0.40 MPa (T_{sat} = 8.93°C)				p = 5.0 bar = 0.50 MPa (T_{sat} = 15.74°C)			
Sat.	0.05089	231.97	252.32	0.9145	0.04086	235.64	256.07	0.9117
10	0.05119	232.87	253.35	0.9182				
20	0.05397	241.37	262.96	0.9515	0.04188	239.40	260.34	0.9264
30	0.05662	249.89	272.54	0.9837	0.04416	248.20	270.28	0.9597
40	0.05917	258.47	282.14	1.0148	0.04633	256.99	280.16	0.9918
50	0.06164	267.13	291.79	1.0452	0.04842	265.83	290.04	1.0229
60	0.06405	275.89	301.51	1.0748	0.05043	274.73	299.95	1.0531
70	0.06641	284.75	311.32	1.1038	0.05240	283.72	309.92	1.0825
80	0.06873	293.73	321.23	1.1322	0.05432	292.80	319.96	1.1114
90	0.07102	302.84	331.25	1.1602	0.05620	302.00	330.10	1.1397
100	0.07327	312.07	341.38	1.1878	0.05805	311.31	340.33	1.1675
110	0.07550	321.44	351.64	1.2149	0.05988	320.74	350.68	1.1949
120	0.07771	330.94	362.03	1.2417	0.06168	330.30	361.14	1.2218
130	0.07991	340.58	372.54	1.2681	0.06347	339.98	371.72	1.2484
140	0.08208	350.35	383.18	1.2941	0.06524	349.79	382.42	1.2746
	p = 6.0 bar = 0.60 MPa (T_{sat} = 21.58°C)				p = 7.0 bar = 0.70 MPa (T_{sat} = 26.72°C)			
Sat.	0.03408	238.74	259.19	0.9097	0.02918	241.42	261.85	0.9080
30	0.03581	246.41	267.89	0.9388	0.02979	244.51	265.37	0.9197
40	0.03774	255.45	278.09	0.9719	0.03157	253.83	275.93	0.9539
50	0.03958	264.48	288.23	1.0037	0.03324	263.08	286.35	0.9867
60	0.04134	273.54	298.35	1.0346	0.03482	272.31	296.69	1.0182
70	0.04304	282.66	308.48	1.0645	0.03634	281.57	307.01	1.0487
80	0.04469	291.86	318.67	1.0938	0.03781	290.88	317.35	1.0784
90	0.04631	301.14	328.93	1.1225	0.03924	300.27	327.74	1.1074
100	0.04790	310.53	339.27	1.1505	0.04064	309.74	338.19	1.1358
110	0.04946	320.03	349.70	1.1781	0.04201	319.31	348.71	1.1637
120	0.05099	329.64	360.24	1.2053	0.04335	328.98	359.33	1.1910
130	0.05251	339.38	370.88	1.2320	0.04468	338.76	370.04	1.2179
140	0.05402	349.23	381.64	1.2584	0.04599	348.66	380.86	1.2444
150	0.05550	359.21	392.52	1.2844	0.04729	358.68	391.79	1.2706
160	0.05698	369.32	403.51	1.3100	0.04857	368.82	402.82	1.2963

(Continued)

T °C	v m³/kg	u kJ/kg	h kJ/kg	s kJ/kg·K	v m³/kg	u kJ/kg	h kJ/kg	s kJ/kg·K
	$p = 8.0$ bar $= 0.80$ MPa ($T_{sat} = 31.33°C$)				$p = 9.0$ bar $= 0.90$ MPa ($T_{sat} = 35.53°C$)			
Sat.	0.02547	243.78	264.15	0.9066	0.02255	245.88	266.18	0.9054
40	0.02691	252.13	273.66	0.9374	0.02325	250.32	271.25	0.9217
50	0.02846	261.62	284.39	0.9711	0.02472	260.09	282.34	0.9566
60	0.02992	271.04	294.98	1.0034	0.02609	269.72	293.21	0.9897
70	0.03131	280.45	305.50	1.0345	0.02738	279.30	303.94	1.0214
80	0.03264	289.89	316.00	1.0647	0.02861	288.87	314.62	1.0521
90	0.03393	299.37	326.52	1.0940	0.02980	298.46	325.28	1.0819
100	0.03519	308.93	337.08	1.1227	0.03095	308.11	335.96	1.1109
110	0.03642	318.57	347.71	1.1508	0.03207	317.82	346.68	1.1392
120	0.03762	328.31	358.40	1.1784	0.03316	327.62	357.47	1.1670
130	0.03881	338.14	369.19	1.2055	0.03423	337.52	368.33	1.1943
140	0.03997	348.09	380.07	1.2321	0.03529	347.51	379.27	1.2211
150	0.04113	358.15	391.05	1.2584	0.03633	357.61	390.31	1.2475
160	0.04227	368.32	402.14	1.2843	0.03736	367.82	401.44	1.2735
170	0.04340	378.61	413.33	1.3098	0.03838	378.14	412.68	1.2992
180	0.04452	389.02	424.63	1.3351	0.03939	388.57	424.02	1.3245
	$p = 10.0$ bar $= 1.00$ MPa ($T_{sat} = 39.39°C$)				$p = 12.0$ bar $= 1.20$ MPa ($T_{sat} = 46.32°C$)			
Sat.	0.02020	247.77	267.97	0.9043	0.01663	251.03	270.99	0.9023
40	0.02029	248.39	268.68	0.9066				
50	0.02171	258.48	280.19	0.9428	0.01712	254.98	275.52	0.9164
60	0.02301	268.35	291.36	0.9768	0.01835	265.42	287.44	0.9527
70	0.02423	278.11	302.34	1.0093	0.01947	275.59	298.96	0.9868
80	0.02538	287.82	313.20	1.0405	0.02051	285.62	310.24	1.0192
90	0.02649	297.53	324.01	1.0707	0.02150	295.59	321.39	1.0503
100	0.02755	307.27	334.82	1.1000	0.02244	305.54	332.47	1.0804
110	0.02858	317.06	345.65	1.1286	0.02335	315.50	343.52	1.1096
120	0.02959	326.93	356.52	1.1567	0.02423	325.51	354.58	1.1381
130	0.03058	336.88	367.46	1.1841	0.02508	335.58	365.68	1.1660
140	0.03154	346.92	378.46	1.2111	0.02592	345.73	376.83	1.1933
150	0.03250	357.06	389.56	1.2376	0.02674	355.95	388.04	1.2201
160	0.03344	367.31	400.74	1.2638	0.02754	366.27	399.33	1.2465
170	0.03436	377.66	412.02	1.2895	0.02834	376.69	410.70	1.2724
180	0.03528	388.12	423.40	1.3149	0.02912	387.21	422.16	1.2980
	$p = 14.0$ bar $= 1.40$ MPa ($T_{sat} = 52.43°C$)				$p = 16.0$ bar $= 1.60$ MPa ($T_{sat} = 57.92°C$)			
Sat.	0.01405	253.74	273.40	0.9003	0.01208	256.00	275.33	0.8982
60	0.01495	262.17	283.10	0.9297	0.01233	258.48	278.20	0.9069
70	0.01603	272.87	295.31	0.9658	0.01340	269.89	291.33	0.9457
80	0.01701	283.29	307.10	0.9997	0.01435	280.78	303.74	0.9813
90	0.01792	293.55	318.63	1.0319	0.01521	291.39	315.72	1.0148
100	0.01878	303.73	330.02	1.0628	0.01601	301.84	327.46	1.0467
110	0.01960	313.88	341.32	1.0927	0.01677	312.20	339.04	1.0773
120	0.02039	324.05	352.59	1.1218	0.01750	322.53	350.53	1.1069
130	0.02115	334.25	363.86	1.1501	0.01820	332.87	361.99	1.1357
140	0.02189	344.50	375.15	1.1777	0.01887	343.24	373.44	1.1638
150	0.02262	354.82	386.49	1.2048	0.01953	353.66	384.91	1.1912
160	0.02333	365.22	397.89	1.2315	0.02017	364.15	396.43	1.2181
170	0.02403	375.71	409.36	1.2576	0.02080	374.71	407.99	1.2445
180	0.02472	386.29	420.90	1.2834	0.02142	385.35	419.62	1.2704
190	0.02541	396.96	432.53	1.3088	0.02203	396.08	431.33	1.2960
200	0.02608	407.73	444.24	1.3338	0.02263	406.90	443.11	1.3212

R-134a

TABLE A-13

Pressure Conversions:
1 bar = 0.1 MPa
= 10² kPa

Properties of Saturated Ammonia (Liquid–Vapor): Temperature Table

Temp. °C	Press. bar	Specific Volume m³/kg Sat. Liquid $v_f \times 10^3$	Sat. Vapor v_g	Internal Energy kJ/kg Sat. Liquid u_f	Sat. Vapor u_g	Enthalpy kJ/kg Sat. Liquid h_f	Evap. h_{fg}	Sat. Vapor h_g	Entropy kJ/kg·K Sat. Liquid s_f	Sat. Vapor s_g	Temp. °C
−50	0.4086	1.4245	2.6265	−43.94	1264.99	−43.88	1416.20	1372.32	−0.1922	6.1543	−50
−45	0.5453	1.4367	2.0060	−22.03	1271.19	−21.95	1402.52	1380.57	−0.0951	6.0523	−45
−40	0.7174	1.4493	1.5524	−0.10	1277.20	0.00	1388.56	1388.56	0.0000	5.9557	−40
−36	0.8850	1.4597	1.2757	17.47	1281.87	17.60	1377.17	1394.77	0.0747	5.8819	−36
−32	1.0832	1.4703	1.0561	35.09	1286.41	35.25	1365.55	1400.81	0.1484	5.8111	−32
−30	1.1950	1.4757	0.9634	43.93	1288.63	44.10	1359.65	1403.75	0.1849	5.7767	−30
−28	1.3159	1.4812	0.8803	52.78	1290.82	52.97	1353.68	1406.66	0.2212	5.7430	−28
−26	1.4465	1.4867	0.8056	61.65	1292.97	61.86	1347.65	1409.51	0.2572	5.7100	−26
−22	1.7390	1.4980	0.6780	79.46	1297.18	79.72	1335.36	1415.08	0.3287	5.6457	−22
−20	1.9019	1.5038	0.6233	88.40	1299.23	88.68	1329.10	1417.79	0.3642	5.6144	−20
−18	2.0769	1.5096	0.5739	97.36	1301.25	97.68	1322.77	1420.45	0.3994	5.5837	−18
−16	2.2644	1.5155	0.5291	106.36	1303.23	106.70	1316.35	1423.05	0.4346	5.5536	−16
−14	2.4652	1.5215	0.4885	115.37	1305.17	115.75	1309.86	1425.61	0.4695	5.5239	−14
−12	2.6798	1.5276	0.4516	124.42	1307.08	124.83	1303.28	1428.11	0.5043	5.4948	−12
−10	2.9089	1.5338	0.4180	133.50	1308.95	133.94	1296.61	1430.55	0.5389	5.4662	−10
−8	3.1532	1.5400	0.3874	142.60	1310.78	143.09	1289.86	1432.95	0.5734	5.4380	−8
−6	3.4134	1.5464	0.3595	151.74	1312.57	152.26	1283.02	1435.28	0.6077	5.4103	−6
−4	3.6901	1.5528	0.3340	160.88	1314.32	161.46	1276.10	1437.56	0.6418	5.3831	−4
−2	3.9842	1.5594	0.3106	170.07	1316.04	170.69	1269.08	1439.78	0.6759	5.3562	−2
0	4.2962	1.5660	0.2892	179.29	1317.71	179.96	1261.97	1441.94	0.7097	5.3298	0
2	4.6270	1.5727	0.2695	188.53	1319.34	189.26	1254.77	1444.03	0.7435	5.3038	2
4	4.9773	1.5796	0.2514	197.80	1320.92	198.59	1247.48	1446.07	0.7770	5.2781	4
6	5.3479	1.5866	0.2348	207.10	1322.47	207.95	1240.09	1448.04	0.8105	5.2529	6
8	5.7395	1.5936	0.2195	216.42	1323.96	217.34	1232.61	1449.94	0.8438	5.2279	8
10	6.1529	1.6008	0.2054	225.77	1325.42	226.75	1225.03	1451.78	0.8769	5.2033	10
12	6.5890	1.6081	0.1923	235.14	1326.82	236.20	1217.35	1453.55	0.9099	5.1791	12
16	7.5324	1.6231	0.1691	253.95	1329.48	255.18	1201.70	1456.87	0.9755	5.1314	16
20	8.5762	1.6386	0.1492	272.86	1331.94	274.26	1185.64	1459.90	1.0404	5.0849	20
24	9.7274	1.6547	0.1320	291.84	1334.19	293.45	1169.16	1462.61	1.1048	5.0394	24
28	10.993	1.6714	0.1172	310.92	1336.20	312.75	1152.24	1465.00	1.1686	4.9948	28
32	12.380	1.6887	0.1043	330.07	1337.97	332.17	1134.87	1467.03	1.2319	4.9509	32
36	13.896	1.7068	0.0930	349.32	1339.47	351.69	1117.00	1468.70	1.2946	4.9078	36
40	15.549	1.7256	0.0831	368.67	1340.70	371.35	1098.62	1469.97	1.3569	4.8652	40
45	17.819	1.7503	0.0725	393.01	1341.81	396.13	1074.84	1470.96	1.4341	4.8125	45
50	20.331	1.7765	0.0634	417.56	1342.42	421.17	1050.09	1471.26	1.5109	4.7604	50

v_f = (table value)/1000

TABLE A-14

Properties of Saturated Ammonia (Liquid–Vapor): Pressure Table

Pressure Conversions:
1 bar = 0.1 MPa
= 10^2 kPa

Press. bar	Temp. °C	Specific Volume m^3/kg Sat. Liquid $v_f \times 10^3$	Sat. Vapor v_g	Internal Energy kJ/kg Sat. Liquid u_f	Sat. Vapor u_g	Enthalpy kJ/kg Sat. Liquid h_f	Evap. h_{fg}	Sat. Vapor h_g	Entropy kJ/kg·K Sat. Liquid s_f	Sat. Vapor s_g	Press. bar
0.40	−50.36	1.4236	2.6795	−45.52	1264.54	−45.46	1417.18	1371.72	−0.1992	6.1618	0.40
0.50	−46.53	1.4330	2.1752	−28.73	1269.31	−28.66	1406.73	1378.07	−0.1245	6.0829	0.50
0.60	−43.28	1.4410	1.8345	−14.51	1273.27	−14.42	1397.76	1383.34	−0.0622	6.0186	0.60
0.70	−40.46	1.4482	1.5884	−2.11	1276.66	−2.01	1389.85	1387.84	−0.0086	5.9643	0.70
0.80	−37.94	1.4546	1.4020	8.93	1279.61	9.04	1382.73	1391.78	0.0386	5.9174	0.80
0.90	−35.67	1.4605	1.2559	18.91	1282.24	19.04	1376.23	1395.27	0.0808	5.8760	0.90
1.00	−33.60	1.4660	1.1381	28.03	1284.61	28.18	1370.23	1398.41	0.1191	5.8391	1.00
1.25	−29.07	1.4782	0.9237	48.03	1289.65	48.22	1356.89	1405.11	0.2018	5.7610	1.25
1.50	−25.22	1.4889	0.7787	65.10	1293.80	65.32	1345.28	1410.61	0.2712	5.6973	1.50
1.75	−21.86	1.4984	0.6740	80.08	1297.33	80.35	1334.92	1415.27	0.3312	5.6435	1.75
2.00	−18.86	1.5071	0.5946	93.50	1300.39	93.80	1325.51	1419.31	0.3843	5.5969	2.00
2.25	−16.15	1.5151	0.5323	105.68	1303.08	106.03	1316.83	1422.86	0.4319	5.5558	2.25
2.50	−13.67	1.5225	0.4821	116.88	1305.49	117.26	1308.76	1426.03	0.4753	5.5190	2.50
2.75	−11.37	1.5295	0.4408	127.26	1307.67	127.68	1301.20	1428.88	0.5152	5.4858	2.75
3.00	−9.24	1.5361	0.4061	136.96	1309.65	137.42	1294.05	1431.47	0.5520	5.4554	3.00
3.25	−7.24	1.5424	0.3765	146.06	1311.46	146.57	1287.27	1433.84	0.5864	5.4275	3.25
3.50	−5.36	1.5484	0.3511	154.66	1313.14	155.20	1280.81	1436.01	0.6186	5.4016	3.50
3.75	−3.58	1.5542	0.3289	162.80	1314.68	163.38	1274.64	1438.03	0.6489	5.3774	3.75
4.00	−1.90	1.5597	0.3094	170.55	1316.12	171.18	1268.71	1439.89	0.6776	5.3548	4.00
4.25	−0.29	1.5650	0.2921	177.96	1317.47	178.62	1263.01	1441.63	0.7048	5.3336	4.25
4.50	1.25	1.5702	0.2767	185.04	1318.73	185.75	1257.50	1443.25	0.7308	5.3135	4.50
4.75	2.72	1.5752	0.2629	191.84	1319.91	192.59	1252.18	1444.77	0.7555	5.2946	4.75
5.00	4.13	1.5800	0.2503	198.39	1321.02	199.18	1247.02	1446.19	0.7791	5.2765	5.00
5.25	5.48	1.5847	0.2390	204.69	1322.07	205.52	1242.01	1447.53	0.8018	5.2594	5.25
5.50	6.79	1.5893	0.2286	210.78	1323.06	211.65	1237.15	1448.80	0.8236	5.2430	5.50
5.75	8.05	1.5938	0.2191	216.66	1324.00	217.58	1232.41	1449.99	0.8446	5.2273	5.75
6.00	9.27	1.5982	0.2104	222.37	1324.89	223.32	1227.79	1451.12	0.8649	5.2122	6.00
7.00	13.79	1.6148	0.1815	243.56	1328.04	244.69	1210.38	1455.07	0.9394	5.1576	7.00
8.00	17.84	1.6302	0.1596	262.64	1330.64	263.95	1194.36	1458.30	1.0054	5.1099	8.00
9.00	21.52	1.6446	0.1424	280.05	1332.82	281.53	1179.44	1460.97	1.0649	5.0675	9.00
10.00	24.89	1.6584	0.1285	296.10	1334.66	297.76	1165.42	1463.18	1.1191	5.0294	10.00
12.00	30.94	1.6841	0.1075	324.99	1337.52	327.01	1139.52	1466.53	1.2152	4.9625	12.00
14.00	36.26	1.7080	0.0923	350.58	1339.56	352.97	1115.82	1468.79	1.2987	4.9050	14.00
16.00	41.03	1.7306	0.0808	373.69	1340.97	376.46	1093.77	1470.23	1.3729	4.8542	16.00
18.00	45.38	1.7522	0.0717	394.85	1341.88	398.00	1073.01	1471.01	1.4399	4.8086	18.00
20.00	49.37	1.7731	0.0644	414.44	1342.37	417.99	1053.27	1471.26	1.5012	4.7670	20.00

v_f = (table value)/1000

Ammonia

TABLE A-15

Properties of Superheated Ammonia Vapor

Pressure Conversions:
1 bar = 0.1 MPa
= 10^2 kPa

T °C	v m³/kg	u kJ/kg	h kJ/kg	s kJ/kg·K	v m³/kg	u kJ/kg	h kJ/kg	s kJ/kg·K
	\multicolumn{4}{}{p = 0.4 bar = 0.04 MPa (T_{sat} = −50.36°C)}		p = 0.6 bar = 0.06 MPa (T_{sat} = −43.28°C)			
Sat.	2.6795	1264.54	1371.72	6.1618	1.8345	1273.27	1383.34	6.0186
−50	2.6841	1265.11	1372.48	6.1652				
−45	2.7481	1273.05	1382.98	6.2118				
−40	2.8118	1281.01	1393.48	6.2573	1.8630	1278.62	1390.40	6.0490
−35	2.8753	1288.96	1403.98	6.3018	1.9061	1286.75	1401.12	6.0946
−30	2.9385	1296.93	1414.47	6.3455	1.9491	1294.88	1411.83	6.1390
−25	3.0015	1304.90	1424.96	6.3882	1.9918	1303.01	1422.52	6.1826
−20	3.0644	1312.88	1435.46	6.4300	2.0343	1311.13	1433.19	6.2251
−15	3.1271	1320.87	1445.95	6.4711	2.0766	1319.25	1443.85	6.2668
−10	3.1896	1328.87	1456.45	6.5114	2.1188	1327.37	1454.50	6.3077
−5	3.2520	1336.88	1466.95	6.5509	2.1609	1335.49	1465.14	6.3478
0	3.3142	1344.90	1477.47	6.5898	2.2028	1343.61	1475.78	6.3871
5	3.3764	1352.95	1488.00	6.6280	2.2446	1351.75	1486.43	6.4257
	\multicolumn{4}{}{p = 0.8 bar = 0.08 MPa (T_{sat} = −37.94°C)}		p = 1.0 bar = 0.10 MPa (T_{sat} = −33.60°C)			
Sat.	1.4021	1279.61	1391.78	5.9174	1.1381	1284.61	1398.41	5.8391
−35	1.4215	1284.51	1398.23	5.9446				
−30	1.4543	1292.81	1409.15	5.9900	1.1573	1290.71	1406.44	5.8723
−25	1.4868	1301.09	1420.04	6.0343	1.1838	1299.15	1417.53	5.9175
−20	1.5192	1309.36	1430.90	6.0777	1.2101	1307.57	1428.58	5.9616
−15	1.5514	1317.61	1441.72	6.1200	1.2362	1315.96	1439.58	6.0046
−10	1.5834	1325.85	1452.53	6.1615	1.2621	1324.33	1450.54	6.0467
−5	1.6153	1334.09	1463.31	6.2021	1.2880	1332.67	1461.47	6.0878
0	1.6471	1342.31	1474.08	6.2419	1.3136	1341.00	1472.37	6.1281
5	1.6788	1350.54	1484.84	6.2809	1.3392	1349.33	1483.25	6.1676
10	1.7103	1358.77	1495.60	6.3192	1.3647	1357.64	1494.11	6.2063
15	1.7418	1367.01	1506.35	6.3568	1.3900	1365.95	1504.96	6.2442
20	1.7732	1375.25	1517.10	6.3939	1.4153	1374.27	1515.80	6.2816
	\multicolumn{4}{}{p = 1.5 bar = 0.15 MPa (T_{sat} = −25.22°C)}		p = 2.0 bar = 0.20 MPa (T_{sat} = −18.86°C)			
Sat.	0.7787	1293.80	1410.61	5.6973	0.59460	1300.39	1419.31	5.5969
−25	0.7795	1294.20	1411.13	5.6994				
−20	0.7978	1303.00	1422.67	5.7454				
−15	0.8158	1311.75	1434.12	5.7902	0.60542	1307.43	1428.51	5.6328
−10	0.8336	1320.44	1445.49	5.8338	0.61926	1316.46	1440.31	5.6781
−5	0.8514	1329.08	1456.79	5.8764	0.63294	1325.41	1452.00	5.7221
0	0.8689	1337.68	1468.02	5.9179	0.64648	1334.29	1463.59	5.7649
5	0.8864	1346.25	1479.20	5.9585	0.65989	1343.11	1475.09	5.8066
10	0.9037	1354.78	1490.34	5.9981	0.67320	1351.87	1486.51	5.8473
15	0.9210	1363.29	1501.44	6.0370	0.68640	1360.59	1497.87	5.8871
20	0.9382	1371.79	1512.51	6.0751	0.69952	1369.28	1509.18	5.9260
25	0.9553	1380.28	1523.56	6.1125	0.71256	1377.93	1520.44	5.9641
30	0.9723	1388.76	1534.60	6.1492	0.72553	1386.56	1531.67	6.0014

Ammonia

TABLE A-15

(Continued)

Ammonia

T °C	v m³/kg	u kJ/kg	h kJ/kg	s kJ/kg·K	v m³/kg	u kJ/kg	h kJ/kg	s kJ/kg·K
	$p = 2.5$ bar $= 0.25$ MPa ($T_{sat} = -13.67°C$)				$p = 3.0$ bar $= 0.30$ MPa ($T_{sat} = -9.24°C$)			
Sat.	0.48213	1305.49	1426.03	5.5190	0.40607	1309.65	1431.47	5.4554
−10	0.49051	1312.37	1435.00	5.5534				
−5	0.50180	1321.65	1447.10	5.5989	0.41428	1317.80	1442.08	5.4953
0	0.51293	1330.83	1459.06	5.6431	0.42382	1327.28	1454.43	5.5409
5	0.52393	1339.91	1470.89	5.6860	0.43323	1336.64	1466.61	5.5851
10	0.53482	1348.91	1482.61	5.7278	0.44251	1345.89	1478.65	5.6280
15	0.54560	1357.84	1494.25	5.7685	0.45169	1355.05	1490.56	5.6697
20	0.55630	1366.72	1505.80	5.8083	0.46078	1364.13	1502.36	5.7103
25	0.56691	1375.55	1517.28	5.8471	0.46978	1373.14	1514.07	5.7499
30	0.57745	1384.34	1528.70	5.8851	0.47870	1382.09	1525.70	5.7886
35	0.58793	1393.10	1540.08	5.9223	0.48756	1391.00	1537.26	5.8264
40	0.59835	1401.84	1551.42	5.9589	0.49637	1399.86	1548.77	5.8635
45	0.60872	1410.56	1562.74	5.9947	0.50512	1408.70	1560.24	5.8998
	$p = 3.5$ bar $= 0.35$ MPa ($T_{sat} = -5.36°C$)				$p = 4.0$ bar $= 0.40$ MPa ($T_{sat} = -1.90°C$)			
Sat.	0.35108	1313.14	1436.01	5.4016	0.30942	1316.12	1439.89	5.3548
0	0.36011	1323.66	1449.70	5.4522	0.31227	1319.95	1444.86	5.3731
10	0.37654	1342.82	1474.61	5.5417	0.32701	1339.68	1470.49	5.4652
20	0.39251	1361.49	1498.87	5.6259	0.34129	1358.81	1495.33	5.5515
30	0.40814	1379.81	1522.66	5.7057	0.35520	1377.49	1519.57	5.6328
40	0.42350	1397.87	1546.09	5.7818	0.36884	1395.85	1543.38	5.7101
60	0.45363	1433.55	1592.32	5.9249	0.39550	1431.97	1590.17	5.8549
80	0.48320	1469.06	1638.18	6.0586	0.42160	1467.77	1636.41	5.9897
100	0.51240	1504.73	1684.07	6.1850	0.44733	1503.64	1682.58	6.1169
120	0.54136	1540.79	1730.26	6.3056	0.47280	1539.85	1728.97	6.2380
140	0.57013	1577.38	1776.92	6.4213	0.49808	1576.55	1775.79	6.3541
160	0.59876	1614.60	1824.16	6.5330	0.52323	1613.86	1823.16	6.4661
180	0.62728	1652.51	1872.06	6.6411	0.54827	1651.85	1871.16	6.5744
200	0.65572	1691.15	1920.65	6.7460	0.57322	1690.56	1919.85	6.6796
	$p = 4.5$ bar $= 0.45$ MPa ($T_{sat} = 1.25°C$)				$p = 5.0$ bar $= 0.50$ MPa ($T_{sat} = 4.13°C$)			
Sat.	0.27671	1318.73	1443.25	5.3135	0.25034	1321.02	1446.19	5.2765
10	0.28846	1336.48	1466.29	5.3962	0.25757	1333.22	1462.00	5.3330
20	0.30142	1356.09	1491.72	5.4845	0.26949	1353.32	1488.06	5.4234
30	0.31401	1375.15	1516.45	5.5674	0.28103	1372.76	1513.28	5.5080
40	0.32631	1393.80	1540.64	5.6460	0.29227	1391.74	1537.87	5.5878
60	0.35029	1430.37	1588.00	5.7926	0.31410	1428.76	1585.81	5.7362
80	0.37369	1466.47	1634.63	5.9285	0.33535	1465.16	1632.84	5.8733
100	0.39671	1502.55	1681.07	6.0564	0.35621	1501.46	1679.56	6.0020
120	0.41947	1538.91	1727.67	6.1781	0.37681	1537.97	1726.37	6.1242
140	0.44205	1575.73	1774.65	6.2946	0.39722	1574.90	1773.51	6.2412
160	0.46448	1613.13	1822.15	6.4069	0.41749	1612.40	1821.14	6.3537
180	0.48681	1651.20	1870.26	6.5155	0.43765	1650.54	1869.36	6.4626
200	0.50905	1689.97	1919.04	6.6208	0.45771	1689.38	1918.24	6.5681

TABLE A-15

(Continued)

Pressure Conversions:
1 bar = 0.1 MPa
= 10² kPa

T °C	v m³/kg	u kJ/kg	h kJ/kg	s kJ/kg·K	v m³/kg	u kJ/kg	h kJ/kg	s kJ/kg·K
	$p = 5.5$ bar $= 0.55$ MPa ($T_{sat} = 6.79°C$)				$p = 6.0$ bar $= 0.60$ MPa ($T_{sat} = 9.27°C$)			
Sat.	0.22861	1323.06	1448.80	5.2430	0.21038	1324.89	1451.12	5.2122
10	0.23227	1329.88	1457.63	5.2743	0.21115	1326.47	1453.16	5.2195
20	0.24335	1350.50	1484.34	5.3671	0.22155	1347.62	1480.55	5.3145
30	0.25403	1370.35	1510.07	5.4534	0.23152	1367.90	1506.81	5.4026
40	0.26441	1389.64	1535.07	5.5345	0.24118	1387.52	1532.23	5.4851
50	0.27454	1408.53	1559.53	5.6114	0.25059	1406.67	1557.03	5.5631
60	0.28449	1427.13	1583.60	5.6848	0.25981	1425.49	1581.38	5.6373
80	0.30398	1463.85	1631.04	5.8230	0.27783	1462.52	1629.22	5.7768
100	0.32307	1500.36	1678.05	5.9525	0.29546	1499.25	1676.52	5.9071
120	0.34190	1537.02	1725.07	6.0753	0.31281	1536.07	1723.76	6.0304
140	0.36054	1574.07	1772.37	6.1926	0.32997	1573.24	1771.22	6.1481
160	0.37903	1611.66	1820.13	6.3055	0.34699	1610.92	1819.12	6.2613
180	0.39742	1649.88	1868.46	6.4146	0.36390	1649.22	1867.56	6.3707
200	0.41571	1688.79	1917.43	6.5203	0.38071	1688.20	1916.63	6.4766
	$p = 7.0$ bar $= 0.70$ MPa ($T_{sat} = 13.79°C$)				$p = 8.0$ bar $= 0.80$ MPa ($T_{sat} = 17.84°C$)			
Sat.	0.18148	1328.04	1455.07	5.1576	0.15958	1330.64	1458.30	5.1099
20	0.18721	1341.72	1472.77	5.2186	0.16138	1335.59	1464.70	5.1318
30	0.19610	1362.88	1500.15	5.3104	0.16948	1357.71	1493.29	5.2277
40	0.20464	1383.20	1526.45	5.3958	0.17720	1378.77	1520.53	5.3161
50	0.21293	1402.90	1551.95	5.4760	0.18465	1399.05	1546.77	5.3986
60	0.22101	1422.16	1576.87	5.5519	0.19189	1418.77	1572.28	5.4763
80	0.23674	1459.85	1625.56	5.6939	0.20590	1457.14	1621.86	5.6209
100	0.25205	1497.02	1673.46	5.8258	0.21949	1494.77	1670.37	5.7545
120	0.26709	1534.16	1721.12	5.9502	0.23280	1532.24	1718.48	5.8801
140	0.28193	1571.57	1768.92	6.0688	0.24590	1569.89	1766.61	5.9995
160	0.29663	1609.44	1817.08	6.1826	0.25886	1607.96	1815.04	6.1140
180	0.31121	1647.90	1865.75	6.2925	0.27170	1646.57	1863.94	6.2243
200	0.32571	1687.02	1915.01	6.3988	0.28445	1685.83	1913.39	6.3311
	$p = 9.0$ bar $= 0.90$ MPa ($T_{sat} = 21.52°C$)				$p = 10.0$ bar $= 1.00$ MPa ($T_{sat} = 24.89°C$)			
Sat.	0.14239	1332.82	1460.97	5.0675	0.12852	1334.66	1463.18	5.0294
30	0.14872	1352.36	1486.20	5.1520	0.13206	1346.82	1478.88	5.0816
40	0.15582	1374.21	1514.45	5.2436	0.13868	1369.52	1508.20	5.1768
50	0.16263	1395.11	1541.47	5.3286	0.14499	1391.07	1536.06	5.2644
60	0.16922	1415.32	1567.61	5.4083	0.15106	1411.79	1562.86	5.3460
80	0.18191	1454.39	1618.11	5.5555	0.16270	1451.60	1614.31	5.4960
100	0.19416	1492.50	1667.24	5.6908	0.17389	1490.20	1664.10	5.6332
120	0.20612	1530.30	1715.81	5.8176	0.18478	1528.35	1713.13	5.7612
140	0.21788	1568.20	1764.29	5.9379	0.19545	1566.51	1761.96	5.8823
160	0.22948	1606.46	1813.00	6.0530	0.20598	1604.97	1810.94	5.9981
180	0.24097	1645.24	1862.12	6.1639	0.21638	1643.91	1860.29	6.1095
200	0.25237	1684.64	1911.77	6.2711	0.22670	1683.44	1910.14	6.2171

Ammonia

TABLE A-15

(Continued)

Ammonia

T °C	v m³/kg	u kJ/kg	h kJ/kg	s kJ/kg·K	v m³/kg	u kJ/kg	h kJ/kg	s kJ/kg·K
	p = 12.0 bar = 1.20 MPa (T_{sat} = 30.94°C)				p = 14.0 bar = 1.40 MPa (T_{sat} = 36.26°C)			
Sat.	0.10751	1337.52	1466.53	4.9625	0.09231	1339.56	1468.79	4.9050
40	0.11287	1359.73	1495.18	5.0553	0.09432	1349.29	1481.33	4.9453
60	0.12378	1404.54	1553.07	5.2347	0.10423	1396.97	1542.89	5.1360
80	0.13387	1445.91	1606.56	5.3906	0.11324	1440.06	1598.59	5.2984
100	0.14347	1485.55	1657.71	5.5315	0.12172	1480.79	1651.20	5.4433
120	0.15275	1524.41	1707.71	5.6620	0.12986	1520.41	1702.21	5.5765
140	0.16181	1563.09	1757.26	5.7850	0.13777	1559.63	1752.52	5.7013
160	0.17072	1601.95	1806.81	5.9021	0.14552	1598.92	1802.65	5.8198
180	0.17950	1641.23	1856.63	6.0145	0.15315	1638.53	1852.94	5.9333
200	0.18819	1681.05	1906.87	6.1230	0.16068	1678.64	1903.59	6.0427
220	0.19680	1721.50	1957.66	6.2282	0.16813	1719.35	1954.73	6.1485
240	0.20534	1762.63	2009.04	6.3303	0.17551	1760.72	2006.43	6.2513
260	0.21382	1804.48	2061.06	6.4297	0.18283	1802.78	2058.75	6.3513
280	0.22225	1847.04	2113.74	6.5267	0.19010	1845.55	2111.69	6.4488
	p = 16.0 bar = 1.60 MPa (T_{sat} = 41.03°C)				p = 18.0 bar = 1.80 MPa (T_{sat} = 45.38°C)			
Sat.	0.08079	1340.97	1470.23	4.8542	0.07174	1341.88	1471.01	4.8086
60	0.08951	1389.06	1532.28	5.0461	0.07801	1380.77	1521.19	4.9627
80	0.09774	1434.02	1590.40	5.2156	0.08565	1427.79	1581.97	5.1399
100	0.10539	1475.93	1644.56	5.3648	0.09267	1470.97	1637.78	5.2937
120	0.11268	1516.34	1696.64	5.5008	0.09931	1512.22	1690.98	5.4326
140	0.11974	1556.14	1747.72	5.6276	0.10570	1552.61	1742.88	5.5614
160	0.12663	1595.85	1798.45	5.7475	0.11192	1592.76	1794.23	5.6828
180	0.13339	1635.81	1849.23	5.8621	0.11801	1633.08	1845.50	5.7985
200	0.14005	1676.21	1900.29	5.9723	0.12400	1673.78	1896.98	5.9096
220	0.14663	1717.18	1951.79	6.0789	0.12991	1715.00	1948.83	6.0170
240	0.15314	1758.79	2003.81	6.1823	0.13574	1756.85	2001.18	6.1210
260	0.15959	1801.07	2056.42	6.2829	0.14152	1799.35	2054.08	6.2222
280	0.16599	1844.05	2109.64	6.3809	0.14724	1842.55	2107.58	6.3207
	p = 20.0 bar = 2.00 MPa (T_{sat} = 49.37°C)							
Sat.	0.06445	1342.37	1471.26	4.7670				
60	0.06875	1372.05	1509.54	4.8838				
80	0.07596	1421.36	1573.27	5.0696				
100	0.08248	1465.89	1630.86	5.2283				
120	0.08861	1508.03	1685.24	5.3703				
140	0.09447	1549.03	1737.98	5.5012				
160	0.10016	1589.65	1789.97	5.6241				
180	0.10571	1630.32	1841.74	5.7409				
200	0.11116	1671.33	1893.64	5.8530				
220	0.11652	1712.82	1945.87	5.9611				
240	0.12182	1754.90	1998.54	6.0658				
260	0.12706	1797.63	2051.74	6.1675				
280	0.13224	1841.03	2105.50	6.2665				

TABLE A-16

Properties of Saturated Propane (Liquid–Vapor): Temperature Table

Pressure Conversions:
1 bar = 0.1 MPa
= 10² kPa

v_f = (table value)/1000

Temp. °C	Press. bar	Specific Volume m³/kg Sat. Liquid $v_f \times 10^3$	Sat. Vapor v_g	Internal Energy kJ/kg Sat. Liquid u_f	Sat. Vapor u_g	Enthalpy kJ/kg Sat. Liquid h_f	Evap. h_{fg}	Sat. Vapor h_g	Entropy kJ/kg·K Sat. Liquid s_f	Sat. Vapor s_g	Temp. °C
−100	0.02888	1.553	11.27	−128.4	319.5	−128.4	480.4	352.0	−0.634	2.140	−100
−90	0.06426	1.578	5.345	−107.8	329.3	−107.8	471.4	363.6	−0.519	2.055	−90
−80	0.1301	1.605	2.774	−87.0	339.3	−87.0	462.4	375.4	−0.408	1.986	−80
−70	0.2434	1.633	1.551	−65.8	349.5	−65.8	453.1	387.3	−0.301	1.929	−70
−60	0.4261	1.663	0.9234	−44.4	359.9	−44.3	443.5	399.2	−0.198	1.883	−60
−50	0.7046	1.694	0.5793	−22.5	370.4	−22.4	433.6	411.2	−0.098	1.845	−50
−40	1.110	1.728	0.3798	−0.2	381.0	0.0	423.2	423.2	0.000	1.815	−40
−30	1.677	1.763	0.2585	22.6	391.6	22.9	412.1	435.0	0.096	1.791	−30
−20	2.444	1.802	0.1815	45.9	402.4	46.3	400.5	446.8	0.190	1.772	−20
−10	3.451	1.844	0.1309	69.8	413.2	70.4	388.0	458.4	0.282	1.757	−10
0	4.743	1.890	0.09653	94.2	423.8	95.1	374.5	469.6	0.374	1.745	0
4	5.349	1.910	0.08591	104.2	428.1	105.3	368.8	474.1	0.410	1.741	4
8	6.011	1.931	0.07666	114.3	432.3	115.5	362.9	478.4	0.446	1.737	8
12	6.732	1.952	0.06858	124.6	436.5	125.9	356.8	482.7	0.482	1.734	12
16	7.515	1.975	0.06149	135.0	440.7	136.4	350.5	486.9	0.519	1.731	16
20	8.362	1.999	0.05525	145.4	444.8	147.1	343.9	491.0	0.555	1.728	20
24	9.278	2.024	0.04973	156.1	448.9	158.0	337.0	495.0	0.591	1.725	24
28	10.27	2.050	0.04483	166.9	452.9	169.0	329.9	498.9	0.627	1.722	28
32	11.33	2.078	0.04048	177.8	456.7	180.2	322.4	502.6	0.663	1.720	32
36	12.47	2.108	0.03659	188.9	460.6	191.6	314.6	506.2	0.699	1.717	36
40	13.69	2.140	0.03310	200.2	464.3	203.1	306.5	509.6	0.736	1.715	40
44	15.00	2.174	0.02997	211.7	467.9	214.9	298.0	512.9	0.772	1.712	44
48	16.40	2.211	0.02714	223.4	471.4	227.0	288.9	515.9	0.809	1.709	48
52	17.89	2.250	0.02459	235.3	474.6	239.3	279.3	518.6	0.846	1.705	52
56	19.47	2.293	0.02227	247.4	477.7	251.9	269.2	521.1	0.884	1.701	56
60	21.16	2.340	0.02015	259.8	480.6	264.8	258.4	523.2	0.921	1.697	60
65	23.42	2.406	0.01776	275.7	483.6	281.4	243.8	525.2	0.969	1.690	65
70	25.86	2.483	0.01560	292.3	486.1	298.7	227.7	526.4	1.018	1.682	70
75	28.49	2.573	0.01363	309.5	487.8	316.8	209.8	526.6	1.069	1.671	75
80	31.31	2.683	0.01182	327.6	488.2	336.0	189.2	525.2	1.122	1.657	80
85	34.36	2.827	0.01011	347.2	486.9	356.9	164.7	521.6	1.178	1.638	85
90	37.64	3.038	0.008415	369.4	482.2	380.8	133.1	513.9	1.242	1.608	90
95	41.19	3.488	0.006395	399.8	467.4	414.2	79.5	493.7	1.330	1.546	95
96.7	42.48	4.535	0.004535	434.9	434.9	454.2	0.0	454.2	1.437	1.437	96.7

Propane

TABLE A-17

Properties of Saturated Propane (Liquid–Vapor): Pressure Table

Pressure Conversions:
1 bar = 0.1 MPa
= 10² kPa

Press. bar	Temp. °C	Specific Volume m³/kg Sat. Liquid $v_f \times 10^3$	Sat. Vapor v_g	Internal Energy kJ/kg Sat. Liquid u_f	Sat. Vapor u_g	Enthalpy kJ/kg Sat. Liquid h_f	Evap. h_{fg}	Sat. Vapor h_g	Entropy kJ/kg·K Sat. Liquid s_f	Sat. Vapor s_g	Press. bar
0.05	−93.28	1.570	6.752	−114.6	326.0	−114.6	474.4	359.8	−0.556	2.081	0.05
0.10	−83.87	1.594	3.542	−95.1	335.4	−95.1	465.9	370.8	−0.450	2.011	0.10
0.25	−69.55	1.634	1.513	−64.9	350.0	−64.9	452.7	387.8	−0.297	1.927	0.25
0.50	−56.93	1.672	0.7962	−37.7	363.1	−37.6	440.5	402.9	−0.167	1.871	0.50
0.75	−48.68	1.698	0.5467	−19.6	371.8	−19.5	432.3	412.8	−0.085	1.841	0.75
1.00	−42.38	1.719	0.4185	−5.6	378.5	−5.4	425.7	420.3	−0.023	1.822	1.00
2.00	−25.43	1.781	0.2192	33.1	396.6	33.5	406.9	440.4	0.139	1.782	2.00
3.00	−14.16	1.826	0.1496	59.8	408.7	60.3	393.3	453.6	0.244	1.762	3.00
4.00	−5.46	1.865	0.1137	80.8	418.0	81.5	382.0	463.5	0.324	1.751	4.00
5.00	1.74	1.899	0.09172	98.6	425.7	99.5	372.1	471.6	0.389	1.743	5.00
6.00	7.93	1.931	0.07680	114.2	432.2	115.3	363.0	478.3	0.446	1.737	6.00
7.00	13.41	1.960	0.06598	128.2	438.0	129.6	354.6	484.2	0.495	1.733	7.00
8.00	18.33	1.989	0.05776	141.0	443.1	142.6	346.7	489.3	0.540	1.729	8.00
9.00	22.82	2.016	0.05129	152.9	447.6	154.7	339.1	493.8	0.580	1.726	9.00
10.00	26.95	2.043	0.04606	164.0	451.8	166.1	331.8	497.9	0.618	1.723	10.00
11.00	30.80	2.070	0.04174	174.5	455.6	176.8	324.7	501.5	0.652	1.721	11.00
12.00	34.39	2.096	0.03810	184.4	459.1	187.0	317.8	504.8	0.685	1.718	12.00
13.00	37.77	2.122	0.03499	193.9	462.2	196.7	311.0	507.7	0.716	1.716	13.00
14.00	40.97	2.148	0.03231	203.0	465.2	206.0	304.4	510.4	0.745	1.714	14.00
15.00	44.01	2.174	0.02997	211.7	467.9	215.0	297.9	512.9	0.772	1.712	15.00
16.00	46.89	2.200	0.02790	220.1	470.4	223.6	291.4	515.0	0.799	1.710	16.00
17.00	49.65	2.227	0.02606	228.3	472.7	232.0	285.0	517.0	0.824	1.707	17.00
18.00	52.30	2.253	0.02441	236.2	474.9	240.2	278.6	518.8	0.849	1.705	18.00
19.00	54.83	2.280	0.02292	243.8	476.9	248.2	272.2	520.4	0.873	1.703	19.00
20.00	57.27	2.308	0.02157	251.3	478.7	255.9	265.9	521.8	0.896	1.700	20.00
22.00	61.90	2.364	0.01921	265.8	481.7	271.0	253.0	524.0	0.939	1.695	22.00
24.00	66.21	2.424	0.01721	279.7	484.3	285.5	240.1	525.6	0.981	1.688	24.00
26.00	70.27	2.487	0.01549	293.1	486.2	299.6	226.9	526.5	1.021	1.681	26.00
28.00	74.10	2.555	0.01398	306.2	487.5	313.4	213.2	526.6	1.060	1.673	28.00
30.00	77.72	2.630	0.01263	319.2	488.1	327.1	198.9	526.0	1.097	1.664	30.00
35.00	86.01	2.862	0.009771	351.4	486.3	361.4	159.1	520.5	1.190	1.633	35.00
40.00	93.38	3.279	0.007151	387.9	474.7	401.0	102.3	503.3	1.295	1.574	40.00
42.48	96.70	4.535	0.004535	434.9	434.9	454.2	0.0	454.2	1.437	1.437	42.48

$v_f = $ (table value)/1000

Propane

TABLE A-18

Properties of Superheated Propane Vapor

T °C	v m³/kg	u kJ/kg	h kJ/kg	s kJ/kg · K	v m³/kg	u kJ/kg	h kJ/kg	s kJ/kg · K
	p = 0.05 bar = 0.005 MPa (T_sat = −93.28°C)				p = 0.1 bar = 0.01 MPa (T_sat = −83.87°C)			
Sat.	6.752	326.0	359.8	2.081	3.542	367.3	370.8	2.011
−90	6.877	329.4	363.8	2.103				
−80	7.258	339.8	376.1	2.169	3.617	339.5	375.7	2.037
−70	7.639	350.6	388.8	2.233	3.808	350.3	388.4	2.101
−60	8.018	361.8	401.9	2.296	3.999	361.5	401.5	2.164
−50	8.397	373.3	415.3	2.357	4.190	373.1	415.0	2.226
−40	8.776	385.1	429.0	2.418	4.380	385.0	428.8	2.286
−30	9.155	397.4	443.2	2.477	4.570	397.3	443.0	2.346
−20	9.533	410.1	457.8	2.536	4.760	410.0	457.6	2.405
−10	9.911	423.2	472.8	2.594	4.950	423.1	472.6	2.463
0	10.29	436.8	488.2	2.652	5.139	436.7	488.1	2.520
10	10.67	450.8	504.1	2.709	5.329	450.6	503.9	2.578
20	11.05	270.6	520.4	2.765	5.518	465.1	520.3	2.634
	p = 0.5 bar = 0.05 MPa (T_sat = −56.93°C)				p = 1.0 bar = 0.1 MPa (T_sat = −42.38°C)			
Sat.	0.796	363.1	402.9	1.871	0.4185	378.5	420.3	1.822
−50	0.824	371.3	412.5	1.914				
−40	0.863	383.4	426.6	1.976	0.4234	381.5	423.8	1.837
−30	0.903	396.0	441.1	2.037	0.4439	394.2	438.6	1.899
−20	0.942	408.8	455.9	2.096	0.4641	407.3	453.7	1.960
−10	0.981	422.1	471.1	2.155	0.4842	420.7	469.1	2.019
0	1.019	435.8	486.7	2.213	0.5040	434.4	484.8	2.078
10	1.058	449.8	502.7	2.271	0.5238	448.6	501.0	2.136
20	1.096	464.3	519.1	2.328	0.5434	463.3	517.6	2.194
30	1.135	479.2	535.9	2.384	0.5629	478.2	534.5	2.251
40	1.173	494.6	553.2	2.440	0.5824	493.7	551.9	2.307
50	1.211	510.4	570.9	2.496	0.6018	509.5	569.7	2.363
60	1.249	526.7	589.1	2.551	0.6211	525.8	587.9	2.419
	p = 2.0 bar = 0.2 MPa (T_sat = −25.43°C)				p = 3.0 bar = 0.3 MPa (T_sat = −14.16°C)			
Sat.	0.2192	396.6	440.4	1.782	0.1496	408.7	453.6	1.762
−20	0.2251	404.0	449.0	1.816				
−10	0.2358	417.7	464.9	1.877	0.1527	414.7	460.5	1.789
0	0.2463	431.8	481.1	1.938	0.1602	429.0	477.1	1.851
10	0.2566	446.3	497.6	1.997	0.1674	443.8	494.0	1.912
20	0.2669	461.1	514.5	2.056	0.1746	458.8	511.2	1.971
30	0.2770	476.3	531.7	2.113	0.1816	474.2	528.7	2.030
40	0.2871	491.9	549.3	2.170	0.1885	490.1	546.6	2.088
50	0.2970	507.9	567.3	2.227	0.1954	506.2	564.8	2.145
60	0.3070	524.3	585.7	2.283	0.2022	522.7	583.4	2.202
70	0.3169	541.1	604.5	2.339	0.2090	539.6	602.3	2.258
80	0.3267	558.4	623.7	2.394	0.2157	557.0	621.7	2.314
90	0.3365	576.1	643.4	2.449	0.2223	574.8	641.5	2.369

Pressure Conversions:
1 bar = 0.1 MPa
= 10² kPa

Propane

TABLE A-18

(Continued)

T °C	v m³/kg	u kJ/kg	h kJ/kg	s kJ/kg·K	v m³/kg	u kJ/kg	h kJ/kg	s kJ/kg·K
	$p = 4.0$ bar $= 0.4$ MPa ($T_{sat} = -5.46$°C)				$p = 5.0$ bar $= 0.5$ MPa ($T_{sat} = 1.74$°C)			
Sat.	0.1137	418.0	463.5	1.751	0.09172	425.7	471.6	1.743
0	0.1169	426.1	472.9	1.786				
10	0.1227	441.2	490.3	1.848	0.09577	438.4	486.3	1.796
20	0.1283	456.6	507.9	1.909	0.1005	454.1	504.3	1.858
30	0.1338	472.2	525.7	1.969	0.1051	470.0	522.5	1.919
40	0.1392	488.1	543.8	2.027	0.1096	486.1	540.9	1.979
50	0.1445	504.4	562.2	2.085	0.1140	502.5	559.5	2.038
60	0.1498	521.1	581.0	2.143	0.1183	519.4	578.5	2.095
70	0.1550	538.1	600.1	2.199	0.1226	536.6	597.9	2.153
80	0.1601	555.7	619.7	2.255	0.1268	554.1	617.5	2.209
90	0.1652	573.5	639.6	2.311	0.1310	572.1	637.6	2.265
100	0.1703	591.8	659.9	2.366	0.1351	590.5	658.0	2.321
110	0.1754	610.4	680.6	2.421	0.1392	609.3	678.9	2.376
	$p = 6.0$ bar $= 0.6$ MPa ($T_{sat} = 7.93$°C)				$p = 7.0$ bar $= 0.7$ MPa ($T_{sat} = 13.41$°C)			
Sat.	0.07680	432.2	478.3	1.737	0.06598	438.0	484.2	1.733
10	0.07769	435.6	482.2	1.751				
20	0.08187	451.5	500.6	1.815	0.06847	448.8	496.7	1.776
30	0.08588	467.7	519.2	1.877	0.07210	465.2	515.7	1.840
40	0.08978	484.0	537.9	1.938	0.07558	481.9	534.8	1.901
50	0.09357	500.7	556.8	1.997	0.07896	498.7	554.0	1.962
60	0.09729	517.6	576.0	2.056	0.08225	515.9	573.5	2.021
70	0.1009	535.0	595.5	2.113	0.08547	533.4	593.2	2.079
80	0.1045	552.7	615.4	2.170	0.08863	551.2	613.2	2.137
90	0.1081	570.7	635.6	2.227	0.09175	569.4	633.6	2.194
100	0.1116	589.2	656.2	2.283	0.09482	587.9	654.3	2.250
110	0.1151	608.0	677.1	2.338	0.09786	606.8	675.3	2.306
120	0.1185	627.3	698.4	2.393	0.1009	626.2	696.8	2.361
	$p = 8.0$ bar $= 0.8$ MPa ($T_{sat} = 18.33$°C)				$p = 9.0$ bar $= 0.9$ MPa ($T_{sat} = 22.82$°C)			
Sat.	0.05776	443.1	489.3	1.729	0.05129	447.2	493.8	1.726
20	0.05834	445.9	492.6	1.740				
30	0.06170	462.7	512.1	1.806	0.05355	460.0	508.2	1.774
40	0.06489	479.6	531.5	1.869	0.05653	477.2	528.1	1.839
50	0.06796	496.7	551.1	1.930	0.05938	494.7	548.1	1.901
60	0.07094	514.0	570.8	1.990	0.06213	512.2	568.1	1.962
70	0.07385	531.6	590.7	2.049	0.06479	530.0	588.3	2.022
80	0.07669	549.6	611.0	2.107	0.06738	548.1	608.7	2.081
90	0.07948	567.9	631.5	2.165	0.06992	566.5	629.4	2.138
100	0.08222	586.5	652.3	2.221	0.07241	585.2	650.4	2.195
110	0.08493	605.6	673.5	2.277	0.07487	604.3	671.7	2.252
120	0.08761	625.0	695.1	2.333	0.07729	623.7	693.3	2.307
130	0.09026	644.8	717.0	2.388	0.07969	643.6	715.3	2.363
140	0.09289	665.0	739.3	2.442	0.08206	663.8	737.7	2.418

Pressure Conversions:
1 bar = 0.1 MPa
= 10² kPa

Propane

TABLE A-18

(Continued)

Pressure Conversions:
1 bar = 0.1 MPa
= 10² kPa

T	v	u	h	s	v	u	h	s
°C	m³/kg	kJ/kg	kJ/kg	kJ/kg · K	m³/kg	kJ/kg	kJ/kg	kJ/kg · K

T °C	v m³/kg	u kJ/kg	h kJ/kg	s kJ/kg · K	v m³/kg	u kJ/kg	h kJ/kg	s kJ/kg · K
	p = 10.0 bar = 1.0 MPa (T_{sat} = 26.95°C)				p = 12.0 bar = 1.2 MPa (T_{sat} = 34.39°C)			
Sat.	0.04606	451.8	497.9	1.723	0.03810	459.1	504.8	1.718
30	0.04696	457.1	504.1	1.744				
40	0.04980	474.8	524.6	1.810	0.03957	469.4	516.9	1.757
50	0.05248	492.4	544.9	1.874	0.04204	487.8	538.2	1.824
60	0.05505	510.2	565.2	1.936	0.04436	506.1	559.3	1.889
70	0.05752	528.2	585.7	1.997	0.04657	524.4	580.3	1.951
80	0.05992	546.4	606.3	2.056	0.04869	543.1	601.5	2.012
90	0.06226	564.9	627.2	2.114	0.05075	561.8	622.7	2.071
100	0.06456	583.7	648.3	2.172	0.05275	580.9	644.2	2.129
110	0.06681	603.0	669.8	2.228	0.05470	600.4	666.0	2.187
120	0.06903	622.6	691.6	2.284	0.05662	620.1	688.0	2.244
130	0.07122	642.5	713.7	2.340	0.05851	640.1	710.3	2.300
140	0.07338	662.8	736.2	2.395	0.06037	660.6	733.0	2.355
	p = 14.0 bar = 1.4 MPa (T_{sat} = 40.97°C)				p = 16.0 bar = 1.6 MPa (T_{sat} = 46.89°C)			
Sat.	0.03231	465.2	510.4	1.714	0.02790	470.4	515.0	1.710
50	0.03446	482.6	530.8	1.778	0.02861	476.7	522.5	1.733
60	0.03664	501.6	552.9	1.845	0.03075	496.6	545.8	1.804
70	0.03869	520.4	574.6	1.909	0.03270	516.2	568.5	1.871
80	0.04063	539.4	596.3	1.972	0.03453	535.7	590.9	1.935
90	0.04249	558.6	618.1	2.033	0.03626	555.2	613.2	1.997
100	0.04429	577.9	639.9	2.092	0.03792	574.8	635.5	2.058
110	0.04604	597.5	662.0	2.150	0.03952	594.7	657.9	2.117
120	0.04774	617.5	684.3	2.208	0.04107	614.8	680.5	2.176
130	0.04942	637.7	706.9	2.265	0.04259	635.3	703.4	2.233
140	0.05106	658.3	729.8	2.321	0.04407	656.0	726.5	2.290
150	0.05268	679.2	753.0	2.376	0.04553	677.1	749.9	2.346
160	0.05428	700.5	776.5	2.431	0.04696	698.5	773.6	2.401
	p = 18.0 bar = 1.8 MPa (T_{sat} = 52.30°C)				p = 20.0 bar = 2.0 MPa (T_{sat} = 57.27°C)			
Sat.	0.02441	474.9	518.8	1.705	0.02157	478.7	521.8	1.700
60	0.02606	491.1	538.0	1.763	0.02216	484.8	529.1	1.722
70	0.02798	511.4	561.8	1.834	0.02412	506.3	554.5	1.797
80	0.02974	531.6	585.1	1.901	0.02585	527.1	578.8	1.867
90	0.03138	551.5	608.0	1.965	0.02744	547.6	602.5	1.933
100	0.03293	571.5	630.8	2.027	0.02892	568.1	625.9	1.997
110	0.03443	591.7	653.7	2.087	0.03033	588.5	649.2	2.059
120	0.03586	612.1	676.6	2.146	0.03169	609.2	672.6	2.119
130	0.03726	632.7	699.8	2.204	0.03299	630.0	696.0	2.178
140	0.03863	653.6	723.1	2.262	0.03426	651.2	719.7	2.236
150	0.03996	674.8	746.7	2.318	0.03550	672.5	743.5	2.293
160	0.04127	696.3	770.6	2.374	0.03671	694.2	767.6	2.349
170	0.04256	718.2	794.8	2.429	0.03790	716.2	792.0	2.404
180	0.04383	740.4	819.3	2.484	0.03907	738.5	816.6	2.459

Propane

TABLE A-18

(Continued)

T °C	v m³/kg	u kJ/kg	h kJ/kg	s kJ/kg·K	v m³/kg	u kJ/kg	h kJ/kg	s kJ/kg·K
	p = 22.0 bar = 2.2 MPa (T_sat = 61.90°C)				p = 24.0 bar = 2.4 MPa (T_sat = 66.21°C)			
Sat.	0.01921	481.8	524.0	1.695	0.01721	484.3	525.6	1.688
70	0.02086	500.5	546.4	1.761	0.01802	493.7	536.9	1.722
80	0.02261	522.4	572.1	1.834	0.01984	517.0	564.6	1.801
90	0.02417	543.5	596.7	1.903	0.02141	539.0	590.4	1.873
100	0.02561	564.5	620.8	1.969	0.02283	560.6	615.4	1.941
110	0.02697	585.3	644.6	2.032	0.02414	581.9	639.8	2.006
120	0.02826	606.2	668.4	2.093	0.02538	603.2	664.1	2.068
130	0.02949	627.3	692.2	2.153	0.02656	624.6	688.3	2.129
140	0.03069	648.6	716.1	2.211	0.02770	646.0	712.5	2.188
150	0.03185	670.1	740.2	2.269	0.02880	667.8	736.9	2.247
160	0.03298	691.9	764.5	2.326	0.02986	689.7	761.4	2.304
170	0.03409	714.1	789.1	2.382	0.03091	711.9	786.1	2.360
180	0.03517	736.5	813.9	2.437	0.03193	734.5	811.1	2.416
	p = 26.0 bar = 2.6 MPa (T_sat = 70.27°C)				p = 30.0 bar = 3.0 MPa (T_sat = 77.72°C)			
Sat.	0.01549	486.2	526.5	1.681	0.01263	488.2	526.0	1.664
80	0.01742	511.0	556.3	1.767	0.01318	495.4	534.9	1.689
90	0.01903	534.2	583.7	1.844	0.01506	522.8	568.0	1.782
100	0.02045	556.4	609.6	1.914	0.01654	547.2	596.8	1.860
110	0.02174	578.3	634.8	1.981	0.01783	570.4	623.9	1.932
120	0.02294	600.0	659.6	2.045	0.01899	593.0	650.0	1.999
130	0.02408	621.6	684.2	2.106	0.02007	615.4	675.6	2.063
140	0.02516	643.4	708.8	2.167	0.02109	637.7	701.0	2.126
150	0.02621	665.3	733.4	2.226	0.02206	660.1	726.3	2.186
160	0.02723	687.4	758.2	2.283	0.02300	682.6	751.6	2.245
170	0.02821	709.9	783.2	2.340	0.02390	705.4	777.1	2.303
180	0.02918	732.5	808.4	2.397	0.02478	728.3	802.6	2.360
190	0.03012	755.5	833.8	2.452	0.02563	751.5	828.4	2.417
	p = 35.0 bar = 3.5 MPa (T_sat = 86.01°C)				p = 40.0 bar = 4.0 MPa (T_sat = 93.38°C)			
Sat.	0.00977	486.3	520.5	1.633	0.00715	474.7	503.3	1.574
90	0.01086	502.4	540.5	1.688				
100	0.01270	532.9	577.3	1.788	0.00940	512.1	549.7	1.700
110	0.01408	558.9	608.2	1.870	0.01110	544.7	589.1	1.804
120	0.01526	583.4	636.8	1.944	0.01237	572.1	621.6	1.887
130	0.01631	607.0	664.1	2.012	0.01344	597.4	651.2	1.962
140	0.01728	630.2	690.7	2.077	0.01439	621.9	679.5	2.031
150	0.01819	653.3	717.0	2.140	0.01527	645.9	707.0	2.097
160	0.01906	676.4	743.1	2.201	0.01609	669.7	734.1	2.160
170	0.01989	699.6	769.2	2.261	0.01687	693.4	760.9	2.222
180	0.02068	722.9	795.3	2.319	0.01761	717.3	787.7	2.281
190	0.02146	746.5	821.6	2.376	0.01833	741.2	814.5	2.340
200	0.02221	770.3	848.0	2.433	0.01902	765.3	841.4	2.397

Pressure Conversions:
1 bar = 0.1 MPa
     = 10² kPa

Propane

TABLE A-19

Properties of Selected Solids and Liquids: c_p, ρ, and κ

Substance	Specific Heat, c_p (kJ/kg $\cdot$ K)	Density, ρ (kg/m³)	Thermal Conductivity, κ (W/m $\cdot$ K)
Selected Solids, 300K			
Aluminum	0.903	2700	237
Coal, anthracite	1.260	1350	0.26
Copper	0.385	8930	401
Granite	0.775	2630	2.79
Iron	0.447	7870	80.2
Lead	0.129	11300	35.3
Sand	0.800	1520	0.27
Silver	0.235	10500	429
Soil	1.840	2050	0.52
Steel (AISI 302)	0.480	8060	15.1
Tin	0.227	7310	66.6
Building Materials, 300K			
Brick, common	0.835	1920	0.72
Concrete (stone mix)	0.880	2300	1.4
Glass, plate	0.750	2500	1.4
Hardboard, siding	1.170	640	0.094
Limestone	0.810	2320	2.15
Plywood	1.220	545	0.12
Softwoods (fir, pine)	1.380	510	0.12
Insulating Materials, 300K			
Blanket (glass fiber)	—	16	0.046
Cork	1.800	120	0.039
Duct liner (glass fiber, coated)	0.835	32	0.038
Polystyrene (extruded)	1.210	55	0.027
Vermiculite fill (flakes)	0.835	80	0.068
Saturated Liquids			
Ammonia, 300K	4.818	599.8	0.465
Mercury, 300K	0.139	13529	8.540
Refrigerant 22, 300K	1.267	1183.1	0.085
Refrigerant 134a, 300K	1.434	1199.7	0.081
Unused Engine Oil, 300K	1.909	884.1	0.145
Water, 275K	4.211	999.9	0.574
300K	4.179	996.5	0.613
325K	4.182	987.1	0.645
350K	4.195	973.5	0.668
375K	4.220	956.8	0.681
400K	4.256	937.4	0.688

Sources: Drawn from several sources, these data are only representative. Values can vary depending on temperature, purity, moisture content, and other factors.

TABLE A-20

Ideal Gas Specific Heats of Some Common Gases (kJ/kg · K)

Temp. K	c_p	c_v	k	c_p	c_v	k	c_p	c_v	k	Temp. K
	Air			Nitrogen, N_2			Oxygen, O_2			
250	1.003	0.716	1.401	1.039	0.742	1.400	0.913	0.653	1.398	250
300	1.005	0.718	1.400	1.039	0.743	1.400	0.918	0.658	1.395	300
350	1.008	0.721	1.398	1.041	0.744	1.399	0.928	0.668	1.389	350
400	1.013	0.726	1.395	1.044	0.747	1.397	0.941	0.681	1.382	400
450	1.020	0.733	1.391	1.049	0.752	1.395	0.956	0.696	1.373	450
500	1.029	0.742	1.387	1.056	0.759	1.391	0.972	0.712	1.365	500
550	1.040	0.753	1.381	1.065	0.768	1.387	0.988	0.728	1.358	550
600	1.051	0.764	1.376	1.075	0.778	1.382	1.003	0.743	1.350	600
650	1.063	0.776	1.370	1.086	0.789	1.376	1.017	0.758	1.343	650
700	1.075	0.788	1.364	1.098	0.801	1.371	1.031	0.771	1.337	700
750	1.087	0.800	1.359	1.110	0.813	1.365	1.043	0.783	1.332	750
800	1.099	0.812	1.354	1.121	0.825	1.360	1.054	0.794	1.327	800
900	1.121	0.834	1.344	1.145	0.849	1.349	1.074	0.814	1.319	900
1000	1.142	0.855	1.336	1.167	0.870	1.341	1.090	0.830	1.313	1000

Temp. K	c_p	c_v	k	c_p	c_v	k	c_p	c_v	k	Temp. K
	Carbon Dioxide, CO_2			Carbon Monoxide, CO			Hydrogen, H_2			
250	0.791	0.602	1.314	1.039	0.743	1.400	14.051	9.927	1.416	250
300	0.846	0.657	1.288	1.040	0.744	1.399	14.307	10.183	1.405	300
350	0.895	0.706	1.268	1.043	0.746	1.398	14.427	10.302	1.400	350
400	0.939	0.750	1.252	1.047	0.751	1.395	14.476	10.352	1.398	400
450	0.978	0.790	1.239	1.054	0.757	1.392	14.501	10.377	1.398	450
500	1.014	0.825	1.229	1.063	0.767	1.387	14.513	10.389	1.397	500
550	1.046	0.857	1.220	1.075	0.778	1.382	14.530	10.405	1.396	550
600	1.075	0.886	1.213	1.087	0.790	1.376	14.546	10.422	1.396	600
650	1.102	0.913	1.207	1.100	0.803	1.370	14.571	10.447	1.395	650
700	1.126	0.937	1.202	1.113	0.816	1.364	14.604	10.480	1.394	700
750	1.148	0.959	1.197	1.126	0.829	1.358	14.645	10.521	1.392	750
800	1.169	0.980	1.193	1.139	0.842	1.353	14.695	10.570	1.390	800
900	1.204	1.015	1.186	1.163	0.866	1.343	14.822	10.698	1.385	900
1000	1.234	1.045	1.181	1.185	0.888	1.335	14.983	10.859	1.380	1000

Source: Adapted from K. Wark, *Thermodynamics,* 4th ed., McGraw-Hill, New York, 1983, as based on "Tables of Thermal Properties of Gases," NBS Circular 564, 1955.

TABLE A-21

Variation of $\bar{c}_p$ with Temperature for Selected Ideal Gases

$$\frac{\bar{c}_p}{\bar{R}} = \alpha + \beta T + \gamma T^2 + \delta T^3 + \varepsilon T^4$$

T is in K, equations valid from 300 to 1000 K

Gas	α	$\beta \times 10^3$	$\gamma \times 10^6$	$\delta \times 10^9$	$\varepsilon \times 10^{12}$
CO	3.710	−1.619	3.692	−2.032	0.240
CO_2	2.401	8.735	−6.607	2.002	0
H_2	3.057	2.677	−5.810	5.521	−1.812
H_2O	4.070	−1.108	4.152	−2.964	0.807
O_2	3.626	−1.878	7.055	−6.764	2.156
N_2	3.675	−1.208	2.324	−0.632	−0.226
Air	3.653	−1.337	3.294	−1.913	0.2763
SO_2	3.267	5.324	0.684	−5.281	2.559
CH_4	3.826	−3.979	24.558	−22.733	6.963
C_2H_2	1.410	19.057	−24.501	16.391	−4.135
C_2H_4	1.426	11.383	7.989	−16.254	6.749
Monatomic gases[a]	2.5	0	0	0	0

[a]For monatomic gases, such as He, Ne, and Ar, $\bar{c}_p$ is constant over a wide temperature range and is very nearly equal to $5/2\,\bar{R}$.

Source: Adapted from K. Wark, *Thermodynamics,* 4th ed., McGraw-Hill, New York, 1983, as based on NASA SP-273, U.S. Government Printing Office, Washington, DC, 1971.

TABLE A-22

Ideal Gas Properties of Air

$T(\text{K})$, h and $u(\text{kJ/kg})$, $s°$ $(\text{kJ/kg} \cdot \text{K})$

T	h	u	s°	when Δs = 0[1] p_r	v_r	T	h	u	s°	when Δs = 0 p_r	v_r
200	199.97	142.56	1.29559	0.3363	1707.	450	451.80	322.62	2.11161	5.775	223.6
210	209.97	149.69	1.34444	0.3987	1512.	460	462.02	329.97	2.13407	6.245	211.4
220	219.97	156.82	1.39105	0.4690	1346.	470	472.24	337.32	2.15604	6.742	200.1
230	230.02	164.00	1.43557	0.5477	1205.	480	482.49	344.70	2.17760	7.268	189.5
240	240.02	171.13	1.47824	0.6355	1084.	490	492.74	352.08	2.19876	7.824	179.7
250	250.05	178.28	1.51917	0.7329	979.	500	503.02	359.49	2.21952	8.411	170.6
260	260.09	185.45	1.55848	0.8405	887.8	510	513.32	366.92	2.23993	9.031	162.1
270	270.11	192.60	1.59634	0.9590	808.0	520	523.63	374.36	2.25997	9.684	154.1
280	280.13	199.75	1.63279	1.0889	738.0	530	533.98	381.84	2.27967	10.37	146.7
285	285.14	203.33	1.65055	1.1584	706.1	540	544.35	389.34	2.29906	11.10	139.7
290	290.16	206.91	1.66802	1.2311	676.1	550	554.74	396.86	2.31809	11.86	133.1
295	295.17	210.49	1.68515	1.3068	647.9	560	565.17	404.42	2.33685	12.66	127.0
300	300.19	214.07	1.70203	1.3860	621.2	570	575.59	411.97	2.35531	13.50	121.2
305	305.22	217.67	1.71865	1.4686	596.0	580	586.04	419.55	2.37348	14.38	115.7
310	310.24	221.25	1.73498	1.5546	572.3	590	596.52	427.15	2.39140	15.31	110.6
315	315.27	224.85	1.75106	1.6442	549.8	600	607.02	434.78	2.40902	16.28	105.8
320	320.29	228.42	1.76690	1.7375	528.6	610	617.53	442.42	2.42644	17.30	101.2
325	325.31	232.02	1.78249	1.8345	508.4	620	628.07	450.09	2.44356	18.36	96.92
330	330.34	235.61	1.79783	1.9352	489.4	630	638.63	457.78	2.46048	19.84	92.84
340	340.42	242.82	1.82790	2.149	454.1	640	649.22	465.50	2.47716	20.64	88.99
350	350.49	250.02	1.85708	2.379	422.2	650	659.84	473.25	2.49364	21.86	85.34
360	360.58	257.24	1.88543	2.626	393.4	660	670.47	481.01	2.50985	23.13	81.89
370	370.67	264.46	1.91313	2.892	367.2	670	681.14	488.81	2.52589	24.46	78.61
380	380.77	271.69	1.94001	3.176	343.4	680	691.82	496.62	2.54175	25.85	75.50
390	390.88	278.93	1.96633	3.481	321.5	690	702.52	504.45	2.55731	27.29	72.56
400	400.98	286.16	1.99194	3.806	301.6	700	713.27	512.33	2.57277	28.80	69.76
410	411.12	293.43	2.01699	4.153	283.3	710	724.04	520.23	2.58810	30.38	67.07
420	421.26	300.69	2.04142	4.522	266.6	720	734.82	528.14	2.60319	32.02	64.53
430	431.43	307.99	2.06533	4.915	251.1	730	745.62	536.07	2.61803	33.72	62.13
440	441.61	315.30	2.08870	5.332	236.8	740	756.44	544.02	2.63280	35.50	59.82

1. p_r and v_r data for use with Eqs. 6.41 and 6.42, respectively.

TABLE A-22

(Continued)

$T(K)$, h and $u(kJ/kg)$, $s°$ $(kJ/kg \cdot K)$

T	h	u	s°	when $\Delta s = 0$[1]		T	h	u	s°	when $\Delta s = 0$	
				p_r	v_r					p_r	v_r
750	767.29	551.99	2.64737	37.35	57.63	1300	1395.97	1022.82	3.27345	330.9	11.275
760	778.18	560.01	2.66176	39.27	55.54	1320	1419.76	1040.88	3.29160	352.5	10.747
770	789.11	568.07	2.67595	41.31	53.39	1340	1443.60	1058.94	3.30959	375.3	10.247
780	800.03	576.12	2.69013	43.35	51.64	1360	1467.49	1077.10	3.32724	399.1	9.780
790	810.99	584.21	2.70400	45.55	49.86	1380	1491.44	1095.26	3.34474	424.2	9.337
800	821.95	592.30	2.71787	47.75	48.08	1400	1515.42	1113.52	3.36200	450.5	8.919
820	843.98	608.59	2.74504	52.59	44.84	1420	1539.44	1131.77	3.37901	478.0	8.526
840	866.08	624.95	2.77170	57.60	41.85	1440	1563.51	1150.13	3.39586	506.9	8.153
860	888.27	641.40	2.79783	63.09	39.12	1460	1587.63	1168.49	3.41247	537.1	7.801
880	910.56	657.95	2.82344	68.98	36.61	1480	1611.79	1186.95	3.42892	568.8	7.468
900	932.93	674.58	2.84856	75.29	34.31	1500	1635.97	1205.41	3.44516	601.9	7.152
920	955.38	691.28	2.87324	82.05	32.18	1520	1660.23	1223.87	3.46120	636.5	6.854
940	977.92	708.08	2.89748	89.28	30.22	1540	1684.51	1242.43	3.47712	672.8	6.569
960	1000.55	725.02	2.92128	97.00	28.40	1560	1708.82	1260.99	3.49276	710.5	6.301
980	1023.25	741.98	2.94468	105.2	26.73	1580	1733.17	1279.65	3.50829	750.0	6.046
1000	1046.04	758.94	2.96770	114.0	25.17	1600	1757.57	1298.30	3.52364	791.2	5.804
1020	1068.89	776.10	2.99034	123.4	23.72	1620	1782.00	1316.96	3.53879	834.1	5.574
1040	1091.85	793.36	3.01260	133.3	22.39	1640	1806.46	1335.72	3.55381	878.9	5.355
1060	1114.86	810.62	3.03449	143.9	21.14	1660	1830.96	1354.48	3.56867	925.6	5.147
1080	1137.89	827.88	3.05608	155.2	19.98	1680	1855.50	1373.24	3.58335	974.2	4.949
1100	1161.07	845.33	3.07732	167.1	18.896	1700	1880.1	1392.7	3.5979	1025	4.761
1120	1184.28	862.79	3.09825	179.7	17.886	1750	1941.6	1439.8	3.6336	1161	4.328
1140	1207.57	880.35	3.11883	193.1	16.946	1800	2003.3	1487.2	3.6684	1310	3.944
1160	1230.92	897.91	3.13916	207.2	16.064	1850	2065.3	1534.9	3.7023	1475	3.601
1180	1254.34	915.57	3.15916	222.2	15.241	1900	2127.4	1582.6	3.7354	1655	3.295
1200	1277.79	933.33	3.17888	238.0	14.470	1950	2189.7	1630.6	3.7677	1852	3.022
1220	1301.31	951.09	3.19834	254.7	13.747	2000	2252.1	1678.7	3.7994	2068	2.776
1240	1324.93	968.95	3.21751	272.3	13.069	2050	2314.6	1726.8	3.8303	2303	2.555
1260	1348.55	986.90	3.23638	290.8	12.435	2100	2377.4	1775.3	3.8605	2559	2.356
1280	1372.24	1004.76	3.25510	310.4	11.835	2150	2440.3	1823.8	3.8901	2837	2.175
						2200	2503.2	1872.4	3.9191	3138	2.012
						2250	2566.4	1921.3	3.9474	3464	1.864

Source: Table A-22 is based on J. H. Keenan and J. Kaye, *Gas Tables,* Wiley, New York, 1945.

TABLE A-23

Ideal Gas Properties of Selected Gases

Enthalpy $\bar{h}(T)$ and internal energy $\bar{u}(T)$, in kJ/kmol. Absolute entropy at 1 atm $\bar{s}°(T)$, in kJ/kmol · K.

T(K)	Carbon Dioxide, CO_2 ($\bar{h}_f° = -393{,}520$ kJ/kmol)			Carbon Monoxide, CO ($\bar{h}_f° = -110{,}530$ kJ/kmol)			Water Vapor, H_2O ($\bar{h}_f° = -241{,}820$ kJ/kmol)			Oxygen, O_2 ($\bar{h}_f° = 0$ kJ/kmol)			Nitrogen, N_2 ($\bar{h}_f° = 0$ kJ/kmol)			T(K)
	$\bar{h}$	$\bar{u}$	$\bar{s}°$	$\bar{h}$	$\bar{u}$	$\bar{s}°$	$\bar{h}$	$\bar{u}$	$\bar{s}°$	$\bar{h}$	$\bar{u}$	$\bar{s}°$	$\bar{h}$	$\bar{u}$	$\bar{s}°$	
0	0	0	0	0	0	0	0	0	0	0	0	0	0	0	0	0
220	6,601	4,772	202.966	6,391	4,562	188.683	7,295	5,466	178.576	6,404	4,575	196.171	6,391	4,562	182.638	220
230	6,938	5,026	204.464	6,683	4,771	189.980	7,628	5,715	180.054	6,694	4,782	197.461	6,683	4,770	183.938	230
240	7,280	5,285	205.920	6,975	4,979	191.221	7,961	5,965	181.471	6,984	4,989	198.696	6,975	4,979	185.180	240
250	7,627	5,548	207.337	7,266	5,188	192.411	8,294	6,215	182.831	7,275	5,197	199.885	7,266	5,188	186.370	250
260	7,979	5,817	208.717	7,558	5,396	193.554	8,627	6,466	184.139	7,566	5,405	201.027	7,558	5,396	187.514	260
270	8,335	6,091	210.062	7,849	5,604	194.654	8,961	6,716	185.399	7,858	5,613	202.128	7,849	5,604	188.614	270
280	8,697	6,369	211.376	8,140	5,812	195.173	9,296	6,968	186.616	8,150	5,822	203.191	8,141	5,813	189.673	280
290	9,063	6,651	212.660	8,432	6,020	196.735	9,631	7,219	187.791	8,443	6,032	204.218	8,432	6,021	190.695	290
298	9,364	6,885	213.685	8,669	6,190	197.543	9,904	7,425	188.720	8,682	6,203	205.033	8,669	6,190	191.502	298
300	9,431	6,939	213.915	8,723	6,229	197.723	9,966	7,472	188.928	8,736	6,242	205.213	8,723	6,229	191.682	300
310	9,807	7,230	215.146	9,014	6,437	198.678	10,302	7,725	190.030	9,030	6,453	206.177	9,014	6,437	192.638	310
320	10,186	7,526	216.351	9,306	6,645	199.603	10,639	7,978	191.098	9,325	6,664	207.112	9,306	6,645	193.562	320
330	10,570	7,826	217.534	9,597	6,854	200.500	10,976	8,232	192.136	9,620	6,877	208.020	9,597	6,853	194.459	330
340	10,959	8,131	218.694	9,889	7,062	201.371	11,314	8,487	193.144	9,916	7,090	208.904	9,888	7,061	195.328	340
350	11,351	8,439	219.831	10,181	7,271	202.217	11,652	8,742	194.125	10,213	7,303	209.765	10,180	7,270	196.173	350
360	11,748	8,752	220.948	10,473	7,480	203.040	11,992	8,998	195.081	10,511	7,518	210.604	10,471	7,478	196.995	360
370	12,148	9,068	222.044	10,765	7,689	203.842	12,331	9,255	196.012	10,809	7,733	211.423	10,763	7,687	197.794	370
380	12,552	9,392	223.122	11,058	7,899	204.622	12,672	9,513	196.920	11,109	7,949	212.222	11,055	7,895	198.572	380
390	12,960	9,718	224.182	11,351	8,108	205.383	13,014	9,771	197.807	11,409	8,166	213.002	11,347	8,104	199.331	390
400	13,372	10,046	225.225	11,644	8,319	206.125	13,356	10,030	198.673	11,711	8,384	213.765	11,640	8,314	200.071	400
410	13,787	10,378	226.250	11,938	8,529	206.850	13,699	10,290	199.521	12,012	8,603	214.510	11,932	8,523	200.794	410
420	14,206	10,714	227.258	12,232	8,740	207.549	14,043	10,551	200.350	12,314	8,822	215.241	12,225	8,733	201.499	420
430	14,628	11,053	228.252	12,526	8,951	208.252	14,388	10,813	201.160	12,618	9,043	215.955	12,518	8,943	202.189	430
440	15,054	11,393	229.230	12,821	9,163	208.929	14,734	11,075	201.955	12,923	9,264	216.656	12,811	9,153	202.863	440
450	15,483	11,742	230.194	13,116	9,375	209.593	15,080	11,339	202.734	13,228	9,487	217.342	13,105	9,363	203.523	450
460	15,916	12,091	231.144	13,412	9,587	210.243	15,428	11,603	203.497	13,535	9,710	218.016	13,399	9,574	204.170	460
470	16,351	12,444	232.080	13,708	9,800	210.880	15,777	11,869	204.247	13,842	9,935	218.676	13,693	9,786	204.803	470
480	16,791	12,800	233.004	14,005	10,014	211.504	16,126	12,135	204.982	14,151	10,160	219.326	13,988	9,997	205.424	480
490	17,232	13,158	233.916	14,302	10,228	212.117	16,477	12,403	205.705	14,460	10,386	219.963	14,285	10,210	206.033	490
500	17,678	13,521	234.814	14,600	10,443	212.719	16,828	12,671	206.413	14,770	10,614	220.589	14,581	10,423	206.630	500
510	18,126	13,885	235.700	14,898	10,658	213.310	17,181	12,940	207.112	15,082	10,842	221.206	14,876	10,635	207.216	510
520	18,576	14,253	236.575	15,197	10,874	213.890	17,534	13,211	207.799	15,395	11,071	221.812	15,172	10,848	207.792	520
530	19,029	14,622	237.439	15,497	11,090	214.460	17,889	13,482	208.475	15,708	11,301	222.409	15,469	11,062	208.358	530
540	19,485	14,996	238.292	15,797	11,307	215.020	18,245	13,755	209.139	16,022	11,533	222.997	15,766	11,277	208.914	540
550	19,945	15,372	239.135	16,097	11,524	215.572	18,601	14,028	209.795	16,338	11,765	223.576	16,064	11,492	209.461	550
560	20,407	15,751	239.962	16,399	11,743	216.115	18,959	14,303	210.440	16,654	11,998	224.146	16,363	11,707	209.999	560
570	20,870	16,131	240.789	16,701	11,961	216.649	19,318	14,579	211.075	16,971	12,232	224.708	16,662	11,923	210.528	570
580	21,337	16,515	241.602	17,003	12,181	217.175	19,678	14,856	211.702	17,290	12,467	225.262	16,962	12,139	211.049	580
590	21,807	16,902	242.405	17,307	12,401	217.693	20,039	15,134	212.320	17,609	12,703	225.808	17,262	12,356	211.562	590

Table A-23

TABLE A-23

(Continued)

$\bar{h}$ and $\bar{u}$ in kJ/kmol, $\bar{s}°$ in kJ/kmol · K

T(K)	Carbon Dioxide, CO_2 ($\bar{h}_f° = -393{,}520$ kJ/kmol)			Carbon Monoxide, CO ($\bar{h}_f° = -110{,}530$ kJ/kmol)			Water Vapor, H_2O ($\bar{h}_f° = -241{,}820$ kJ/kmol)			Oxygen, O_2 ($\bar{h}_f° = 0$ kJ/kmol)			Nitrogen, N_2 ($\bar{h}_f° = 0$ kJ/kmol)			T(K)
	$\bar{h}$	$\bar{u}$	$\bar{s}°$	$\bar{h}$	$\bar{u}$	$\bar{s}°$	$\bar{h}$	$\bar{u}$	$\bar{s}°$	$\bar{h}$	$\bar{u}$	$\bar{s}°$	$\bar{h}$	$\bar{u}$	$\bar{s}°$	
600	22,280	17,291	243.199	17,611	12,622	218.204	20,402	15,413	212.920	17,929	12,940	226.346	17,563	12,574	212.066	600
610	22,754	17,683	243.983	17,915	12,843	218.708	20,765	15,693	213.529	18,250	13,178	226.877	17,864	12,792	212.564	610
620	23,231	18,076	244.758	18,221	13,066	219.205	21,130	15,975	214.122	18,572	13,417	227.400	18,166	13,011	213.055	620
630	23,709	18,471	245.524	18,527	13,289	219.695	21,495	16,257	214.707	18,895	13,657	227.918	18,468	13,230	213.541	630
640	24,190	18,869	246.282	18,833	13,512	220.179	21,862	16,541	215.285	19,219	13,898	228.429	18,772	13,450	214.018	640
650	24,674	19,270	247.032	19,141	13,736	220.656	22,230	16,826	215.856	19,544	14,140	228.932	19,075	13,671	214.489	650
660	25,160	19,672	247.773	19,449	13,962	221.127	22,600	17,112	216.419	19,870	14,383	229.430	19,380	13,892	214.954	660
670	25,648	20,078	248.507	19,758	14,187	221.592	22,970	17,399	216.976	20,197	14,626	229.920	19,685	14,114	215.413	670
680	26,138	20,484	249.233	20,068	14,414	222.052	23,342	17,688	217.527	20,524	14,871	230.405	19,991	14,337	215.866	680
690	26,631	20,894	249.952	20,378	14,641	222.505	23,714	17,978	218.071	20,854	15,116	230.885	20,297	14,560	216.314	690
700	27,125	21,305	250.663	20,690	14,870	222.953	24,088	18,268	218.610	21,184	15,364	231.358	20,604	14,784	216.756	700
710	27,622	21,719	251.368	21,002	15,099	223.396	24,464	18,561	219.142	21,514	15,611	231.827	20,912	15,008	217.192	710
720	28,121	22,134	252.065	21,315	15,328	223.833	24,840	18,854	219.668	21,845	15,859	232.291	21,220	15,234	217.624	720
730	28,622	22,552	252.755	21,628	15,558	224.265	25,218	19,148	220.189	22,177	16,107	232.748	21,529	15,460	218.059	730
740	29,124	22,972	253.439	21,943	15,789	224.692	25,597	19,444	220.707	22,510	16,357	233.201	21,839	15,686	218.472	740
750	29,629	23,393	254.117	22,258	16,022	225.115	25,977	19,741	221.215	22,844	16,607	233.649	22,149	15,913	218.889	750
760	30,135	23,817	254.787	22,573	16,255	225.533	26,358	20,039	221.720	23,178	16,859	234.091	22,460	16,141	219.301	760
770	30,644	24,242	255.452	22,890	16,488	225.947	26,741	20,339	222.221	23,513	17,111	234.528	22,772	16,370	219.709	770
780	31,154	24,669	256.110	23,208	16,723	226.357	27,125	20,639	222.717	23,850	17,364	234.960	23,085	16,599	220.113	780
790	31,665	25,097	256.762	23,526	16,957	226.762	27,510	20,941	223.207	24,186	17,618	235.387	23,398	16,830	220.512	790
800	32,179	25,527	257.408	23,844	17,193	227.162	27,896	21,245	223.693	24,523	17,872	235.810	23,714	17,061	220.907	800
810	32,694	25,959	258.048	24,164	17,429	227.559	28,284	21,549	224.174	24,861	18,126	236.230	24,027	17,292	221.298	810
820	33,212	26,394	258.682	24,483	17,665	227.952	28,672	21,855	224.651	25,199	18,382	236.644	24,342	17,524	221.684	820
830	33,730	26,829	259.311	24,803	17,902	228.339	29,062	22,162	225.123	25,537	18,637	237.055	24,658	17,757	222.067	830
840	34,251	27,267	259.934	25,124	18,140	228.724	29,454	22,470	225.592	25,877	18,893	237.462	24,974	17,990	222.447	840
850	34,773	27,706	260.551	25,446	18,379	229.106	29,846	22,779	226.057	26,218	19,150	237.864	25,292	18,224	222.822	850
860	35,296	28,125	261.164	25,768	18,617	229.482	30,240	23,090	226.517	26,559	19,408	238.264	25,610	18,459	223.194	860
870	35,821	28,588	261.770	26,091	18,858	229.856	30,635	23,402	226.973	26,899	19,666	238.660	25,928	18,695	223.562	870
880	36,347	29,031	262.371	26,415	19,099	230.227	31,032	23,715	227.426	27,242	19,925	239.051	26,248	18,931	223.927	880
890	36,876	29,476	262.968	26,740	19,341	230.593	31,429	24,029	227.875	27,584	20,185	239.439	26,568	19,168	224.288	890
900	37,405	29,922	263.559	27,066	19,583	230.957	31,828	24,345	228.321	27,928	20,445	239.823	26,890	19,407	224.647	900
910	37,935	30,369	264.146	27,392	19,826	231.317	32,228	24,662	228.763	28,272	20,706	240.203	27,210	19,644	225.002	910
920	38,467	30,818	264.728	27,719	20,070	231.674	32,629	24,980	229.202	28,616	20,967	240.580	27,532	19,883	225.353	920
930	39,000	31,268	265.304	28,046	20,314	232.028	33,032	25,300	229.637	28,960	21,228	240.953	27,854	20,122	225.701	930
940	39,535	31,719	265.877	28,375	20,559	232.379	33,436	25,621	230.070	29,306	21,491	241.323	28,178	20,362	226.047	940
950	40,070	32,171	266.444	28,703	20,805	232.727	33,841	25,943	230.499	29,652	21,754	241.689	28,501	20,603	226.389	950
960	40,607	32,625	267.007	29,033	21,051	233.072	34,247	26,265	230.924	29,999	22,017	242.052	28,826	20,844	226.728	960
970	41,145	33,081	267.566	29,362	21,298	233.413	34,653	26,588	231.347	30,345	22,280	242.411	29,151	21,086	227.064	970
980	41,685	33,537	268.119	29,693	21,545	233.752	35,061	26,913	231.767	30,692	22,544	242.768	29,476	21,328	227.398	980
990	42,226	33,995	268.670	30,024	21,793	234.088	35,472	27,240	232.184	31,041	22,809	243.120	29,803	21,571	227.728	990

(Continued)

$\bar{h}$ and $\bar{u}$ in kJ/kmol, $\bar{s}°$ in kJ/kmol·K

T(K)	Carbon Dioxide, CO_2 ($\bar{h}_f° = -393{,}520$ kJ/kmol)			Carbon Monoxide, CO ($\bar{h}_f° = -110{,}530$ kJ/kmol)			Water Vapor, H_2O ($\bar{h}_f° = -241{,}820$ kJ/kmol)			Oxygen, O_2 ($\bar{h}_f° = 0$ kJ/kmol)			Nitrogen, N_2 ($\bar{h}_f° = 0$ kJ/kmol)			T(K)
	$\bar{h}$	$\bar{u}$	$\bar{s}°$	$\bar{h}$	$\bar{u}$	$\bar{s}°$	$\bar{h}$	$\bar{u}$	$\bar{s}°$	$\bar{h}$	$\bar{u}$	$\bar{s}°$	$\bar{h}$	$\bar{u}$	$\bar{s}°$	
1000	42,769	34,455	269.215	30,355	22,041	234.421	35,882	27,568	232.597	31,389	23,075	243.471	30,129	21,815	228.057	1000
1020	43,859	35,378	270.293	31,020	22,540	235.079	36,709	28,228	233.415	32,088	23,607	244.164	30,784	22,304	228.706	1020
1040	44,953	36,306	271.354	31,688	23,041	235.728	37,542	28,895	234.223	32,789	24,142	244.844	31,442	22,795	229.344	1040
1060	46,051	37,238	272.400	32,357	23,544	236.364	38,380	29,567	235.020	33,490	24,677	245.513	32,101	23,288	229.973	1060
1080	47,153	38,174	273.430	33,029	24,049	236.992	39,223	30,243	235.806	34,194	25,214	246.171	32,762	23,782	230.591	1080
1100	48,258	39,112	274.445	33,702	24,557	237.609	40,071	30,925	236.584	34,899	25,753	246.818	33,426	24,280	231.199	1100
1120	49,369	40,057	275.444	34,377	25,065	238.217	40,923	31,611	237.352	35,606	26,294	247.454	34,092	24,780	231.799	1120
1140	50,484	41,006	276.430	35,054	25,575	238.817	41,780	32,301	238.110	36,314	26,836	248.081	34,760	25,282	232.391	1140
1160	51,602	41,957	277.403	35,733	26,088	239.407	42,642	32,997	238.859	37,023	27,379	248.698	35,430	25,786	232.973	1160
1180	52,724	42,913	278.362	36,406	26,602	239.989	43,509	33,698	239.600	37,734	27,923	249.307	36,104	26,291	233.549	1180
1200	53,848	43,871	279.307	37,095	27,118	240.663	44,380	34,403	240.333	38,447	28,469	249.906	36,777	26,799	234.115	1200
1220	54,977	44,834	280.238	37,780	27,637	241.128	45,256	35,112	241.057	39,162	29,018	250.497	37,452	27,308	234.673	1220
1240	56,108	45,799	281.158	38,466	28,426	241.686	46,137	35,827	241.773	39,877	29,568	251.079	38,129	27,819	235.223	1240
1260	57,244	46,768	282.066	39,154	28,678	242.236	47,022	36,546	242.482	40,594	30,118	251.653	38,807	28,331	235.766	1260
1280	58,381	47,739	282.962	39,884	29,201	242.780	47,912	37,270	243.183	41,312	30,670	252.219	39,488	28,845	236.302	1280
1300	59,522	48,713	283.847	40,534	29,725	243.316	48,807	38,000	243.877	42,033	31,224	252.776	40,170	29,361	236.831	1300
1320	60,666	49,691	284.722	41,266	30,251	243.844	49,707	38,732	244.564	42,753	31,778	253.325	40,853	29,878	237.353	1320
1340	61,813	50,672	285.586	41,919	30,778	244.366	50,612	39,470	245.243	43,475	32,334	253.868	41,539	30,398	237.867	1340
1360	62,963	51,656	286.439	42,613	31,306	244.880	51,521	40,213	245.915	44,198	32,891	254.404	42,227	30,919	238.376	1360
1380	64,116	52,643	287.283	43,309	31,836	245.388	52,434	40,960	246.582	44,923	33,449	254.932	42,915	31,441	238.878	1380
1400	65,271	53,631	288.106	44,007	32,367	245.889	53,351	41,711	247.241	45,648	34,008	255.454	43,605	31,964	239.375	1400
1420	66,427	54,621	288.934	44,707	32,900	246.385	54,273	42,466	247.895	46,374	34,567	255.968	44,295	32,489	239.865	1420
1440	67,586	55,614	289.743	45,408	33,434	246.876	55,198	43,226	248.543	47,102	35,129	256.475	44,988	33,014	240.350	1440
1460	68,748	56,609	290.542	46,110	33,971	247.360	56,128	43,989	249.185	47,831	35,692	256.978	45,682	33,543	240.827	1460
1480	69,911	57,606	291.333	46,813	34,508	247.839	57,062	44,756	249.820	48,561	36,256	257.474	46,377	34,071	241.301	1480
1500	71,078	58,606	292.114	47,517	35,046	248.312	57,999	45,528	250.450	49,292	36,821	257.965	47,073	34,601	241.768	1500
1520	72,246	59,609	292.888	48,222	35,584	248.778	58,942	46,304	251.074	50,024	37,387	258.450	47,771	35,133	242.228	1520
1540	73,417	60,613	293.654	48,928	36,124	249.240	59,888	47,084	251.693	50,756	37,952	258.928	48,470	35,665	242.685	1540
1560	74,590	61,620	294.411	49,635	36,665	249.695	60,838	47,868	252.305	51,490	38,520	259.402	49,168	36,197	243.137	1560
1580	75,763	62,630	295.161	50,344	37,207	250.147	61,792	48,655	252.912	52,224	39,088	259.870	49,869	36,732	243.585	1580
1600	76,944	63,741	295.901	51,053	37,750	250.592	62,748	49,445	253.513	52,961	39,658	260.333	50,571	37,268	244.028	1600
1620	78,123	64,653	296.632	51,763	38,293	251.033	63,709	50,240	254.111	53,696	40,227	260.791	51,275	37,806	244.464	1620
1640	79,303	65,668	297.356	52,472	38,837	251.470	64,675	51,039	254.703	54,434	40,799	261.242	51,980	38,344	244.896	1640
1660	80,486	66,592	298.072	53,184	39,382	251.901	65,643	51,841	255.290	55,172	41,370	261.690	52,686	38,884	245.324	1660
1680	81,670	67,702	298.781	53,895	39,927	252.329	66,614	52,646	255.873	55,912	41,944	262.132	53,393	39,424	245.747	1680
1700	82,856	68,721	299.482	54,609	40,474	252.751	67,589	53,455	256.450	56,652	42,517	262.571	54,099	39,965	246.166	1700
1720	84,043	69,742	300.177	55,323	41,023	253.169	68,567	54,267	257.022	57,394	43,093	263.005	54,807	40,507	246.580	1720
1740	85,231	70,764	300.863	56,039	41,572	253.582	69,550	55,083	257.589	58,136	43,669	263.435	55,516	41,049	246.990	1740

Table A-23

TABLE A-23

(Continued)

$\bar{h}$ and $\bar{u}$ in kJ/kmol, $\bar{s}°$ in kJ/kmol·K

T(K)	Carbon Dioxide, CO_2 ($\bar{h}_f° = -393{,}520$ kJ/kmol)			Carbon Monoxide, CO ($\bar{h}_f° = -110{,}530$ kJ/kmol)			Water Vapor, H_2O ($\bar{h}_f° = -241{,}820$ kJ/kmol)			Oxygen, O_2 ($\bar{h}_f° = 0$ kJ/kmol)			Nitrogen, N_2 ($\bar{h}_f° = 0$ kJ/kmol)			T(K)
	$\bar{h}$	$\bar{u}$	$\bar{s}°$	$\bar{h}$	$\bar{u}$	$\bar{s}°$	$\bar{h}$	$\bar{u}$	$\bar{s}°$	$\bar{h}$	$\bar{u}$	$\bar{s}°$	$\bar{h}$	$\bar{u}$	$\bar{s}°$	
1760	86,420	71,787	301.543	56,756	42,123	253.991	70,535	55,902	258.151	58,800	44,247	263.861	56,227	41,594	247.396	1760
1780	87,612	72,812	302.271	57,473	42,673	254.398	71,523	56,723	258.708	59,624	44,825	264.283	56,938	42,139	247.798	1780
1800	88,806	73,840	302.884	58,191	43,225	254.797	72,513	57,547	259.262	60,371	45,405	264.701	57,651	42,685	248.195	1800
1820	90,000	74,868	303.544	58,910	43,778	255.194	73,507	58,375	259.811	61,118	45,986	265.113	58,363	43,231	248.589	1820
1840	91,196	75,897	304.198	59,629	44,331	255.587	74,506	59,207	260.357	61,866	46,568	265.521	59,075	43,777	248.979	1840
1860	92,394	76,929	304.845	60,351	44,886	255.976	75,506	60,042	260.898	62,616	47,151	265.925	59,790	44,324	249.365	1860
1880	93,593	77,962	305.487	61,071	45,441	256.361	76,511	60,880	261.436	63,365	47,734	266.326	60,504	44,873	249.748	1880
1900	94,793	78,996	306.122	61,794	45,997	256.743	77,517	61,720	261.969	64,116	48,319	266.722	61,220	45,423	250.128	1900
1920	95,995	80,031	306.751	62,516	46,552	257.122	78,527	62,564	262.497	64,868	48,904	267.115	61,936	45,973	250.502	1920
1940	97,197	81,067	307.374	63,238	47,108	257.497	79,540	63,411	263.022	65,620	49,490	267.505	62,654	46,524	250.874	1940
1960	98,401	82,105	307.992	63,961	47,665	257.868	80,555	64,259	263.542	66,374	50,078	267.891	63,381	47,075	251.242	1960
1980	99,606	83,144	308.604	64,684	48,221	258.236	81,573	65,111	264.059	67,127	50,665	268.275	64,090	47,627	251.607	1980
2000	100,804	84,185	309.210	65,408	48,780	258.600	82,593	65,965	264.571	67,881	51,253	268.655	64,810	48,181	251.969	2000
2050	103,835	86,791	310.701	67,224	50,179	259.494	85,156	68,111	265.838	69,772	52,727	269.588	66,612	49,567	252.858	2050
2100	106,864	89,404	312.160	69,044	51,584	260.370	87,735	70,275	267.081	71,668	54,208	270.504	68,417	50,957	253.726	2100
2150	109,898	92,023	313.589	70,864	52,988	261.226	90,330	72,454	268.301	73,573	55,697	271.399	70,226	52,351	254.578	2150
2200	112,939	94,648	314.988	72,688	54,396	262.065	92,940	74,649	269.500	75,484	57,192	272.278	72,040	53,749	255.412	2200
2250	115,984	97,277	316.356	74,516	55,809	262.887	95,562	76,855	270.679	77,397	58,690	273.136	73,856	55,149	256.227	2250
2300	119,035	99,912	317.695	76,345	57,222	263.692	98,199	79,076	271.839	79,316	60,193	273.981	75,676	56,553	257.027	2300
2350	122,091	102,552	319.011	78,178	58,640	264.480	100,846	81,308	272.978	81,243	61,704	274.809	77,496	57,958	257.810	2350
2400	125,152	105,197	320.302	80,015	60,060	265.253	103,508	83,553	274.098	83,174	63,219	275.625	79,320	59,366	258.580	2400
2450	128,219	107,849	321.566	81,852	61,482	266.012	106,183	85,811	275.201	85,112	64,742	276.424	81,149	60,779	259.332	2450
2500	131,290	110,504	322.808	83,692	62,906	266.755	108,868	88,082	276.286	87,057	66,271	277.207	82,981	62,195	260.073	2500
2550	134,368	113,166	324.026	85,537	64,335	267.485	111,565	90,364	277.354	89,004	67,802	277.979	84,814	63,613	260.799	2550
2600	137,449	115,832	325.222	87,383	65,766	268.202	114,273	92,656	278.407	90,956	69,339	278.738	86,650	65,033	261.512	2600
2650	140,533	118,500	326.396	89,230	67,197	268.905	116,991	94,958	279.441	92,916	70,883	279.485	88,488	66,455	262.213	2650
2700	143,620	121,172	327.549	91,077	68,628	269.596	119,717	97,269	280.462	94,881	72,433	280.219	90,328	67,880	262.902	2700
2750	146,713	123,849	328.684	92,930	70,066	270.285	122,453	99,588	281.464	96,852	73,987	280.942	92,171	69,306	263.577	2750
2800	149,808	126,528	329.800	94,784	71,504	270.943	125,198	101,917	282.453	98,826	75,546	281.654	94,014	70,734	264.241	2800
2850	152,908	129,212	330.896	96,639	72,945	271.602	127,952	104,256	283.429	100,808	77,112	282.357	95,859	72,163	264.895	2850
2900	156,009	131,898	331.975	98,495	74,383	272.249	130,717	106,605	284.390	102,793	78,682	283.048	97,705	73,593	265.538	2900
2950	159,117	134,589	333.037	100,352	75,825	272.884	133,486	108,959	285.338	104,785	80,258	283.728	99,556	75,028	266.170	2950
3000	162,226	137,283	334.084	102,210	77,267	273.508	136,264	111,321	286.273	106,780	81,837	284.399	101,407	76,464	266.793	3000
3050	165,341	139,982	335.114	104,073	78,715	274.123	139,051	113,692	287.194	108,778	83,419	285.060	103,260	77,902	267.404	3050
3100	168,456	142,681	336.126	105,939	80,164	274.730	141,846	116,072	288.102	110,784	85,009	285.713	105,115	79,341	268.007	3100
3150	171,576	145,385	337.124	107,802	81,612	275.326	144,648	118,458	288.999	112,795	86,601	286.355	106,972	80,782	268.601	3150
3200	174,695	148,089	338.109	109,667	83,061	275.914	147,457	120,851	289.884	114,809	88,203	286.989	108,830	82,224	269.186	3200
3250	177,822	150,801	339.069	111,534	84,513	276.494	150,272	123,250	290.756	116,827	89,804	287.614	110,690	83,668	269.763	3250

Source: Table A-23 is based on the JANAF Thermochemical Tables, NSRDS-NBS-37, 1971.

TABLE A-24

Constants for the van der Waals, Redlich–Kwong, and Benedict–Webb–Rubin Equations of State

1. van der Waals and Redlich–Kwong: Constants for pressure in bar, specific volume in m^3/kmol, and temperature in K

Substance	van der Waals		Redlich–Kwong	
	a $\;bar\left(\dfrac{m^3}{kmol}\right)^2$	b $\dfrac{m^3}{kmol}$	a $\;bar\left(\dfrac{m^3}{kmol}\right)^2 K^{1/2}$	b $\dfrac{m^3}{kmol}$
Air	1.368	0.0367	15.989	0.02541
Butane (C_4H_{10})	13.86	0.1162	289.55	0.08060
Carbon dioxide (CO_2)	3.647	0.0428	64.43	0.02963
Carbon monoxide (CO)	1.474	0.0395	17.22	0.02737
Methane (CH_4)	2.293	0.0428	32.11	0.02965
Nitrogen (N_2)	1.366	0.0386	15.53	0.02677
Oxygen (O_2)	1.369	0.0317	17.22	0.02197
Propane (C_3H_8)	9.349	0.0901	182.23	0.06242
Refrigerant 12	10.49	0.0971	208.59	0.06731
Sulfur dioxide (SO_2)	6.883	0.0569	144.80	0.03945
Water (H_2O)	5.531	0.0305	142.59	0.02111

Source: Calculated from critical data.

2. Benedict–Webb–Rubin: Constants for pressure in bar, specific volume in m^3/kmol, and temperature in K

Substance	a	A	b	B	c	C	α	γ
C_4H_{10}	1.9073	10.218	0.039998	0.12436	3.206×10^5	1.006×10^6	1.101×10^{-3}	0.0340
CO_2	0.1386	2.7737	0.007210	0.04991	1.512×10^4	1.404×10^5	8.47×10^{-5}	0.00539
CO	0.0371	1.3590	0.002632	0.05454	1.054×10^3	8.676×10^3	1.350×10^{-4}	0.0060
CH_4	0.0501	1.8796	0.003380	0.04260	2.579×10^3	2.287×10^4	1.244×10^{-4}	0.0060
N_2	0.0254	1.0676	0.002328	0.04074	7.381×10^2	8.166×10^3	1.272×10^{-4}	0.0053

Source: H. W. Cooper and J. C. Goldfrank, *Hydrocarbon Processing,* 46 (12): 141 (1967).

TABLE A-25

Thermochemical Properties of Selected Substances at 298K and 1 atm

Substance	Formula	Molar Mass, M (kg/kmol)	Enthalpy of Formation, $\bar{h}_f^\circ$ (kJ/kmol)	Gibbs Function of Formation, $\bar{g}_f^\circ$ (kJ/kmol)	Absolute Entropy, $\bar{s}^\circ$ (kJ/kmol · K)	Heating Values	
						Higher, HHV (kJ/kg)	Lower, LHV (kJ/kg)
Carbon	C(s)	12.01	0	0	5.74	32,770	32,770
Hydrogen	$H_2(g)$	2.016	0	0	130.57	141,780	119,950
Nitrogen	$N_2(g)$	28.01	0	0	191.50	—	—
Oxygen	$O_2(g)$	32.00	0	0	205.03	—	—
Carbon monoxide	CO(g)	28.01	−110,530	−137,150	197.54	—	—
Carbon dioxide	$CO_2(g)$	44.01	−393,520	−394,380	213.69	—	—
Water	$H_2O(g)$	18.02	−241,820	−228,590	188.72	—	—
Water	$H_2O(l)$	18.02	−285,830	−237,180	69.95	—	—
Hydrogen peroxide	$H_2O_2(g)$	34.02	−136,310	−105,600	232.63	—	—
Ammonia	$NH_3(g)$	17.03	−46,190	−16,590	192.33	—	—
Oxygen	O(g)	16.00	249,170	231,770	160.95	—	—
Hydrogen	H(g)	1.008	218,000	203,290	114.61	—	—
Nitrogen	N(g)	14.01	472,680	455,510	153.19	—	—
Hydroxyl	OH(g)	17.01	39,460	34,280	183.75	—	—
Methane	$CH_4(g)$	16.04	−74,850	−50,790	186.16	55,510	50,020
Acetylene	$C_2H_2(g)$	26.04	226,730	209,170	200.85	49,910	48,220
Ethylene	$C_2H_4(g)$	28.05	52,280	68,120	219.83	50,300	47,160
Ethane	$C_2H_6(g)$	30.07	−84,680	−32,890	229.49	51,870	47,480
Propylene	$C_3H_6(g)$	42.08	20,410	62,720	266.94	48,920	45,780
Propane	$C_3H_8(g)$	44.09	−103,850	−23,490	269.91	50,350	46,360
Butane	$C_4H_{10}(g)$	58.12	−126,150	−15,710	310.03	49,500	45,720
Pentane	$C_5H_{12}(g)$	72.15	−146,440	−8,200	348.40	49,010	45,350
Octane	$C_8H_{18}(g)$	114.22	−208,450	17,320	463.67	48,260	44,790
Octane	$C_8H_{18}(l)$	114.22	−249,910	6,610	360.79	47,900	44,430
Benzene	$C_6H_6(g)$	78.11	82,930	129,660	269.20	42,270	40,580
Methanol	$CH_3OH(g)$	32.04	−200,890	−162,140	239.70	23,850	21,110
Methanol	$CH_3OH(l)$	32.04	−238,810	−166,290	126.80	22,670	19,920
Ethanol	$C_2H_5OH(g)$	46.07	−235,310	−168,570	282.59	30,590	27,720
Ethanol	$C_2H_5OH(l)$	46.07	−277,690	−174,890	160.70	29,670	26,800

Source: Based on JANAF Thermochemical Tables, NSRDS-NBS-37, 1971; *Selected Values of Chemical Thermodynamic Properties,* NBS Tech. Note 270-3, 1968; and *API Research Project 44,* Carnegie Press, 1953. Heating values calculated.

Table A-25

TABLE A-26

Standard Molar Chemical Exergy, $\bar{e}^{ch}$ (kJ/kmol), of Selected Substances at 298 K and p_0

Substance	Formula	Model I[a]	Model II[b]
Nitrogen	$N_2(g)$	640	720
Oxygen	$O_2(g)$	3,950	3,970
Carbon dioxide	$CO_2(g)$	14,175	19,870
Water	$H_2O(g)$	8,635	9,500
Water	$H_2O(l)$	45	900
Carbon (graphite)	$C(s)$	404,590	410,260
Hydrogen	$H_2(g)$	235,250	236,100
Sulfur	$S(s)$	598,160	609,600
Carbon monoxide	$CO(g)$	269,410	275,100
Sulfur dioxide	$SO_2(g)$	301,940	313,400
Nitrogen monoxide	$NO(g)$	88,850	88,900
Nitrogen dioxide	$NO_2(g)$	55,565	55,600
Hydrogen sulfide	$H_2S(g)$	799,890	812,000
Ammonia	$NH_3(g)$	336,685	337,900
Methane	$CH_4(g)$	824,350	831,650
Acetylene	$C_2H_2(g)$	—	1,265,800
Ethylene	$C_2H_4(g)$	—	1,361,100
Ethane	$C_2H_6(g)$	1,482,035	1,495,840
Propylene	$C_3H_6(g)$	—	2,003,900
Propane	$C_3H_8(g)$	—	2,154,000
Butane	$C_4H_{10}(g)$	—	2,805,800
Pentane	$C_5H_{12}(g)$	—	3,463,300
Benzene	$C_6H_6(g)$	—	3,303,600
Octane	$C_8H_{18}(l)$	—	5,413,100
Methanol	$CH_3OH(g)$	715,070	722,300
Methanol	$CH_3OH(l)$	710,745	718,000
Ethanol	$C_2H_5OH(g)$	1,348,330	1,363,900
Ethanol	$C_2H_5OH(l)$	1,342,085	1,357,700

[a]J. Ahrendts, "Die Exergie Chemisch Reaktionsfähiger Systeme," *VDI-Forschungsheft*, VDI-Verlag, Dusseldorf, 579, 1977. Also see "Reference States," *Energy—The International Journal, 5*: 667–677, 1980. In Model I, p_0 = 1.019 atm. This model attempts to impose a criterion that the reference environment be in equilibrium. The reference substances are determined assuming restricted chemical equilibrium for nitric acid and nitrates and unrestricted thermodynamic equilibrium for all other chemical components of the atmosphere, the oceans, and a portion of the Earth's crust. The chemical composition of the gas phase of this model approximates the composition of the natural atmosphere.

[b]J. Szargut, D. R. Morris, and F. R. Steward, *Exergy Analysis of Thermal, Chemical, and Metallurgical Processes*, Hemisphere, New York, 1988. In Model II, p_0 = 1.0 atm. In developing this model a reference substance is selected for each chemical element from among substances that contain the element being considered and that are abundantly present in the natural environment, even though the substances are not in completely mutual stable equilibrium. An underlying rationale for this approach is that substances found abundantly in nature have little economic value. On an overall basis, the chemical composition of the exergy reference environment of Model II is closer than Model I to the composition of the natural environment, but the equilibrium criterion is not always satisfied.

TABLE A-27

Logarithms to the Base 10 of the Equilibrium Constant K

Temp. K				$\log_{10} K$					Temp. °R
	$H_2 \rightleftharpoons 2H$	$O_2 \rightleftharpoons 2O$	$N_2 \rightleftharpoons 2N$	$\frac{1}{2}O_2 + \frac{1}{2}N_2 \rightleftharpoons NO$	$H_2O \rightleftharpoons H_2 + \frac{1}{2}O_2$	$H_2O \rightleftharpoons OH + \frac{1}{2}H_2$	$CO_2 \rightleftharpoons CO + \frac{1}{2}O_2$	$CO_2 + H_2 \rightleftharpoons CO + H_2O$	
298	−71.224	−81.208	−159.600	−15.171	−40.048	−46.054	−45.066	−5.018	537
500	−40.316	−45.880	−92.672	−8.783	−22.886	−26.130	−25.025	−2.139	900
1000	−17.292	−19.614	−43.056	−4.062	−10.062	−11.280	−10.221	−0.159	1800
1200	−13.414	−15.208	−34.754	−3.275	−7.899	−8.811	−7.764	+0.135	2160
1400	−10.630	−12.054	−28.812	−2.712	−6.347	−7.021	−6.014	+0.333	2520
1600	−8.532	−9.684	−24.350	−2.290	−5.180	−5.677	−4.706	+0.474	2880
1700	−7.666	−8.706	−22.512	−2.116	−4.699	−5.124	−4.169	+0.530	3060
1800	−6.896	−7.836	−20.874	−1.962	−4.270	−4.613	−3.693	+0.577	3240
1900	−6.204	−7.058	−19.410	−1.823	−3.886	−4.190	−3.267	+0.619	3420
2000	−5.580	−6.356	−18.092	−1.699	−3.540	−3.776	−2.884	+0.656	3600
2100	−5.016	−5.720	−16.898	−1.586	−3.227	−3.434	−2.539	+0.688	3780
2200	−4.502	−5.142	−15.810	−1.484	−2.942	−3.091	−2.226	+0.716	3960
2300	−4.032	−4.614	−14.818	−1.391	−2.682	−2.809	−1.940	+0.742	4140
2400	−3.600	−4.130	−13.908	−1.305	−2.443	−2.520	−1.679	+0.764	4320
2500	−3.202	−3.684	−13.070	−1.227	−2.224	−2.270	−1.440	+0.784	4500
2600	−2.836	−3.272	−12.298	−1.154	−2.021	−2.038	−1.219	+0.802	4680
2700	−2.494	−2.892	−11.580	−1.087	−1.833	−1.823	−1.015	+0.818	4860
2800	−2.178	−2.536	−10.914	−1.025	−1.658	−1.624	−0.825	+0.833	5040
2900	−1.882	−2.206	−10.294	−0.967	−1.495	−1.438	−0.649	+0.846	5220
3000	−1.606	−1.898	−9.716	−0.913	−1.343	−1.265	−0.485	+0.858	5400
3100	−1.348	−1.610	−9.174	−0.863	−1.201	−1.103	−0.332	+0.869	5580
3200	−1.106	−1.340	−8.664	−0.815	−1.067	−0.951	−0.189	+0.878	5760
3300	−0.878	−1.086	−8.186	−0.771	−0.942	−0.809	−0.054	+0.888	5940
3400	−0.664	−0.846	−7.736	−0.729	−0.824	−0.674	+0.071	+0.895	6120
3500	−0.462	−0.620	−7.312	−0.690	−0.712	−0.547	+0.190	+0.902	6300

Source: Based on data from the JANAF Thermochemical Tables, NSRDS-NBS-37, 1971.

Index to Tables in English Units

Table A-1E

TABLE A-1E

Atomic or Molecular Weights and Critical Properties of Some Selected Elements and Compounds

Substance	Chemical Formula	M (lb/lbmol)	T_c (°R)	p_c (atm)	$Z_c = \dfrac{p_c v_c}{RT_c}$
Acetylene	C_2H_2	26.04	556	62	0.274
Air (equivalent)	—	28.97	239	37.2	0.284
Ammonia	NH_3	17.03	730	111.3	0.242
Argon	Ar	39.94	272	47.97	0.290
Benzene	C_6H_6	78.11	1013	48.7	0.274
Butane	C_4H_{10}	58.12	765	37.5	0.274
Carbon	C	12.01	—	—	—
Carbon dioxide	CO_2	44.01	548	72.9	0.276
Carbon monoxide	CO	28.01	239	34.5	0.294
Copper	Cu	63.54	—	—	—
Ethane	C_2H_6	30.07	549	48.2	0.285
Ethanol	C_2H_5OH	46.07	929	63.0	0.249
Ethylene	C_2H_4	28.05	510	50.5	0.270
Helium	He	4.003	9.33	2.26	0.300
Hydrogen	H_2	2.016	59.8	12.8	0.304
Methane	CH_4	16.04	344	45.8	0.290
Methanol	CH_3OH	32.04	924	78.5	0.220
Nitrogen	N_2	28.01	227	33.5	0.291
Octane	C_8H_{18}	114.22	1025	24.6	0.258
Oxygen	O_2	32.00	278	49.8	0.290
Propane	C_3H_8	44.09	666	42.1	0.276
Propylene	C_3H_6	42.08	657	45.6	0.276
Refrigerant 12	CCl_2F_2	120.92	693	40.6	0.278
Refrigerant 22	$CHClF_2$	86.48	665	49.1	0.267
Refrigerant 134a	CF_3CH_2F	102.03	673	40.2	0.260
Sulfur dioxide	SO_2	64.06	775	77.7	0.268
Water	H_2O	18.02	1165	218.0	0.233

Sources: Adapted from *International Critical Tables* and L. C. Nelson and E. F. Obert, Generalized Compressibility Charts, *Chem. Eng.*, *617*: 203 (1954).

TABLE A-2E

Properties of Saturated Water (Liquid–Vapor): Temperature Table

H₂O

Temp. °F	Press. lbf/in.²	Specific Volume ft³/lb		Internal Energy Btu/lb		Enthalpy Btu/lb			Entropy Btu/lb · °R		Temp. °F
		Sat. Liquid v_f	Sat. Vapor v_g	Sat. Liquid u_f	Sat. Vapor u_g	Sat. Liquid h_f	Evap. h_{fg}	Sat. Vapor h_g	Sat. Liquid s_f	Sat. Vapor s_g	
32	0.0886	0.01602	3305	−.01	1021.2	−.01	1075.4	1075.4	−.00003	2.1870	32
35	0.0999	0.01602	2948	2.99	1022.2	3.00	1073.7	1076.7	0.00607	2.1764	35
40	0.1217	0.01602	2445	8.02	1023.9	8.02	1070.9	1078.9	0.01617	2.1592	40
45	0.1475	0.01602	2037	13.04	1025.5	13.04	1068.1	1081.1	0.02618	2.1423	45
50	0.1780	0.01602	1704	18.06	1027.2	18.06	1065.2	1083.3	0.03607	2.1259	50
52	0.1917	0.01603	1589	20.06	1027.8	20.07	1064.1	1084.2	0.04000	2.1195	52
54	0.2064	0.01603	1482	22.07	1028.5	22.07	1063.0	1085.1	0.04391	2.1131	54
56	0.2219	0.01603	1383	24.08	1029.1	24.08	1061.9	1085.9	0.04781	2.1068	56
58	0.2386	0.01603	1292	26.08	1029.8	26.08	1060.7	1086.8	0.05159	2.1005	58
60	0.2563	0.01604	1207	28.08	1030.4	28.08	1059.6	1087.7	0.05555	2.0943	60
62	0.2751	0.01604	1129	30.09	1031.1	30.09	1058.5	1088.6	0.05940	2.0882	62
64	0.2952	0.01604	1056	32.09	1031.8	32.09	1057.3	1089.4	0.06323	2.0821	64
66	0.3165	0.01604	988.4	34.09	1032.4	34.09	1056.2	1090.3	0.06704	2.0761	66
68	0.3391	0.01605	925.8	36.09	1033.1	36.09	1055.1	1091.2	0.07084	2.0701	68
70	0.3632	0.01605	867.7	38.09	1033.7	38.09	1054.0	1092.0	0.07463	2.0642	70
72	0.3887	0.01606	813.7	40.09	1034.4	40.09	1052.8	1092.9	0.07839	2.0584	72
74	0.4158	0.01606	763.5	42.09	1035.0	42.09	1051.7	1093.8	0.08215	2.0526	74
76	0.4446	0.01606	716.8	44.09	1035.7	44.09	1050.6	1094.7	0.08589	2.0469	76
78	0.4750	0.01607	673.3	46.09	1036.3	46.09	1049.4	1095.5	0.08961	2.0412	78
80	0.5073	0.01607	632.8	48.08	1037.0	48.09	1048.3	1096.4	0.09332	2.0356	80
82	0.5414	0.01608	595.0	50.08	1037.6	50.08	1047.2	1097.3	0.09701	2.0300	82
84	0.5776	0.01608	559.8	52.08	1038.3	52.08	1046.0	1098.1	0.1007	2.0245	84
86	0.6158	0.01609	527.0	54.08	1038.9	54.08	1044.9	1099.0	0.1044	2.0190	86
88	0.6562	0.01609	496.3	56.07	1039.6	56.07	1043.8	1099.9	0.1080	2.0136	88
90	0.6988	0.01610	467.7	58.07	1040.2	58.07	1042.7	1100.7	0.1117	2.0083	90
92	0.7439	0.01611	440.9	60.06	1040.9	60.06	1041.5	1101.6	0.1153	2.0030	92
94	0.7914	0.01611	415.9	62.06	1041.5	62.06	1040.4	1102.4	0.1189	1.9977	94
96	0.8416	0.01612	392.4	64.05	1041.2	64.06	1039.2	1103.3	0.1225	1.9925	96
98	0.8945	0.01612	370.5	66.05	1042.8	66.05	1038.1	1104.2	0.1261	1.9874	98
100	0.9503	0.01613	350.0	68.04	1043.5	68.05	1037.0	1105.0	0.1296	1.9822	100
110	1.276	0.01617	265.1	78.02	1046.7	78.02	1031.3	1109.3	0.1473	1.9574	110
120	1.695	0.01621	203.0	87.99	1049.9	88.00	1025.5	1113.5	0.1647	1.9336	120
130	2.225	0.01625	157.2	97.97	1053.0	97.98	1019.8	1117.8	0.1817	1.9109	130
140	2.892	0.01629	122.9	107.95	1056.2	107.96	1014.0	1121.9	0.1985	1.8892	140
150	3.722	0.01634	97.0	117.95	1059.3	117.96	1008.1	1126.1	0.2150	1.8684	150
160	4.745	0.01640	77.2	127.94	1062.3	127.96	1002.2	1130.1	0.2313	1.8484	160
170	5.996	0.01645	62.0	137.95	1065.4	137.97	996.2	1134.2	0.2473	1.8293	170
180	7.515	0.01651	50.2	147.97	1068.3	147.99	990.2	1138.2	0.2631	1.8109	180
190	9.343	0.01657	41.0	158.00	1071.3	158.03	984.1	1142.1	0.2787	1.7932	190
200	11.529	0.01663	33.6	168.04	1074.2	168.07	977.9	1145.9	0.2940	1.7762	200

TABLE A-2E

(Continued)

H₂O

Temp. °F	Press. lbf/in.²	Specific Volume ft³/lb		Internal Energy Btu/lb		Enthalpy Btu/lb			Entropy Btu/lb · °R		Temp. °F
		Sat. Liquid v_f	Sat. Vapor v_g	Sat. Liquid u_f	Sat. Vapor u_g	Sat. Liquid h_f	Evap. h_{fg}	Sat. Vapor h_g	Sat. Liquid s_f	Sat. Vapor s_g	
210	14.13	0.01670	27.82	178.1	1077.0	178.1	971.6	1149.7	0.3091	1.7599	210
212	14.70	0.01672	26.80	180.1	1077.6	180.2	970.3	1150.5	0.3121	1.7567	212
220	17.19	0.01677	23.15	188.2	1079.8	188.2	965.3	1153.5	0.3241	1.7441	220
230	20.78	0.01685	19.39	198.3	1082.6	198.3	958.8	1157.1	0.3388	1.7289	230
240	24.97	0.01692	16.33	208.4	1085.3	208.4	952.3	1160.7	0.3534	1.7143	240
250	29.82	0.01700	13.83	218.5	1087.9	218.6	945.6	1164.2	0.3677	1.7001	250
260	35.42	0.01708	11.77	228.6	1090.5	228.8	938.8	1167.6	0.3819	1.6864	260
270	41.85	0.01717	10.07	238.8	1093.0	239.0	932.0	1170.9	0.3960	1.6731	270
280	49.18	0.01726	8.65	249.0	1095.4	249.2	924.9	1174.1	0.4099	1.6602	280
290	57.53	0.01735	7.47	259.3	1097.7	259.4	917.8	1177.2	0.4236	1.6477	290
300	66.98	0.01745	6.472	269.5	1100.0	269.7	910.4	1180.2	0.4372	1.6356	300
310	77.64	0.01755	5.632	279.8	1102.1	280.1	903.0	1183.0	0.4507	1.6238	310
320	89.60	0.01765	4.919	290.1	1104.2	290.4	895.3	1185.8	0.4640	1.6123	320
330	103.00	0.01776	4.312	300.5	1106.2	300.8	887.5	1188.4	0.4772	1.6010	330
340	117.93	0.01787	3.792	310.9	1108.0	311.3	879.5	1190.8	0.4903	1.5901	340
350	134.53	0.01799	3.346	321.4	1109.8	321.8	871.3	1193.1	0.5033	1.5793	350
360	152.92	0.01811	2.961	331.8	1111.4	332.4	862.9	1195.2	0.5162	1.5688	360
370	173.23	0.01823	2.628	342.4	1112.9	343.0	854.2	1197.2	0.5289	1.5585	370
380	195.60	0.01836	2.339	353.0	1114.3	353.6	845.4	1199.0	0.5416	1.5483	380
390	220.2	0.01850	2.087	363.6	1115.6	364.3	836.2	1200.6	0.5542	1.5383	390
400	247.1	0.01864	1.866	374.3	1116.6	375.1	826.8	1202.0	0.5667	1.5284	400
410	276.5	0.01878	1.673	385.0	1117.6	386.0	817.2	1203.1	0.5792	1.5187	410
420	308.5	0.01894	1.502	395.8	1118.3	396.9	807.2	1204.1	0.5915	1.5091	420
430	343.3	0.01909	1.352	406.7	1118.9	407.9	796.9	1204.8	0.6038	1.4995	430
440	381.2	0.01926	1.219	417.6	1119.3	419.0	786.3	1205.3	0.6161	1.4900	440
450	422.1	0.01943	1.1011	428.6	1119.5	430.2	775.4	1205.6	0.6282	1.4806	450
460	466.3	0.01961	0.9961	439.7	1119.6	441.4	764.1	1205.5	0.6404	1.4712	460
470	514.1	0.01980	0.9025	450.9	1119.4	452.8	752.4	1205.2	0.6525	1.4618	470
480	565.5	0.02000	0.8187	462.2	1118.9	464.3	740.3	1204.6	0.6646	1.4524	480
490	620.7	0.02021	0.7436	473.6	1118.3	475.9	727.8	1203.7	0.6767	1.4430	490
500	680.0	0.02043	0.6761	485.1	1117.4	487.7	714.8	1202.5	0.6888	1.4335	500
520	811.4	0.02091	0.5605	508.5	1114.8	511.7	687.3	1198.9	0.7130	1.4145	520
540	961.5	0.02145	0.4658	532.6	1111.0	536.4	657.5	1193.8	0.7374	1.3950	540
560	1131.8	0.02207	0.3877	548.4	1105.8	562.0	625.0	1187.0	0.7620	1.3749	560
580	1324.3	0.02278	0.3225	583.1	1098.9	588.6	589.3	1178.0	0.7872	1.3540	580
600	1541.0	0.02363	0.2677	609.9	1090.0	616.7	549.7	1166.4	0.8130	1.3317	600
620	1784.4	0.02465	0.2209	638.3	1078.5	646.4	505.0	1151.4	0.8398	1.3075	620
640	2057.1	0.02593	0.1805	668.7	1063.2	678.6	453.4	1131.9	0.8681	1.2803	640
660	2362	0.02767	0.1446	702.3	1042.3	714.4	391.1	1105.5	0.8990	1.2483	660
680	2705	0.03032	0.1113	741.7	1011.0	756.9	309.8	1066.7	0.9350	1.2068	680
700	3090	0.03666	0.0744	801.7	947.7	822.7	167.5	990.2	0.9902	1.1346	700
705.4	3204	0.05053	0.05053	872.6	872.6	902.5	0	902.5	1.0580	1.0580	705.4

Source: Tables A-2E through A-6E are extracted from J. H. Keenan, F. G. Keyes, P. G. Hill, and J. G. Moore, *Steam Tables*, Wiley, New York, 1969.

TABLE A-3E

Properties of Saturated Water (Liquid–Vapor): Pressure Table

Press. lbf/in.^{2}	Temp. °F	Specific Volume ft^3/lb		Internal Energy Btu/lb		Enthalpy Btu/lb			Entropy Btu/lb · °R			Press. lbf/in.^{2}
		Sat. Liquid v_f	Sat. Vapor v_g	Sat. Liquid u_f	Sat. Vapor u_g	Sat. Liquid h_f	Evap. h_{fg}	Sat. Vapor h_g	Sat. Liquid s_f	Evap. s_{fg}	Sat. Vapor s_g	
0.4	72.84	0.01606	792.0	40.94	1034.7	40.94	1052.3	1093.3	0.0800	1.9760	2.0559	0.4
0.6	85.19	0.01609	540.0	53.26	1038.7	53.27	1045.4	1098.6	0.1029	1.9184	2.0213	0.6
0.8	94.35	0.01611	411.7	62.41	1041.7	62.41	1040.2	1102.6	0.1195	1.8773	1.9968	0.8
1.0	101.70	0.01614	333.6	69.74	1044.0	69.74	1036.0	1105.8	0.1327	1.8453	1.9779	1.0
1.2	107.88	0.01616	280.9	75.90	1046.0	75.90	1032.5	1108.4	0.1436	1.8190	1.9626	1.2
1.5	115.65	0.01619	227.7	83.65	1048.5	83.65	1028.0	1111.7	0.1571	1.7867	1.9438	1.5
2.0	126.04	0.01623	173.75	94.02	1051.8	94.02	1022.1	1116.1	0.1750	1.7448	1.9198	2.0
3.0	141.43	0.01630	118.72	109.38	1056.6	109.39	1013.1	1122.5	0.2009	1.6852	1.8861	3.0
4.0	152.93	0.01636	90.64	120.88	1060.2	120.89	1006.4	1127.3	0.2198	1.6426	1.8624	4.0
5.0	162.21	0.01641	73.53	130.15	1063.0	130.17	1000.9	1131.0	0.2349	1.6093	1.8441	5.0
6.0	170.03	0.01645	61.98	137.98	1065.4	138.00	996.2	1134.2	0.2474	1.5819	1.8292	6.0
7.0	176.82	0.01649	53.65	144.78	1067.4	144.80	992.1	1136.9	0.2581	1.5585	1.8167	7.0
8.0	182.84	0.01653	47.35	150.81	1069.2	150.84	988.4	1139.3	0.2675	1.5383	1.8058	8.0
9.0	188.26	0.01656	42.41	156.25	1070.8	156.27	985.1	1141.4	0.2760	1.5203	1.7963	9.0
10	193.19	0.01659	38.42	161.20	1072.2	161.23	982.1	1143.3	0.2836	1.5041	1.7877	10
14.696	211.99	0.01672	26.80	180.10	1077.6	180.15	970.4	1150.5	0.3121	1.4446	1.7567	14.696
15	213.03	0.01672	26.29	181.14	1077.9	181.19	969.7	1150.9	0.3137	1.4414	1.7551	15
20	227.96	0.01683	20.09	196.19	1082.0	196.26	960.1	1156.4	0.3358	1.3962	1.7320	20
25	240.08	0.01692	16.31	208.44	1085.3	208.52	952.2	1160.7	0.3535	1.3607	1.7142	25
30	250.34	0.01700	13.75	218.84	1088.0	218.93	945.4	1164.3	0.3682	1.3314	1.6996	30
35	259.30	0.01708	11.90	227.93	1090.3	228.04	939.3	1167.4	0.3809	1.3064	1.6873	35
40	267.26	0.01715	10.50	236.03	1092.3	236.16	933.8	1170.0	0.3921	1.2845	1.6767	40
45	274.46	0.01721	9.40	243.37	1094.0	243.51	928.8	1172.3	0.4022	1.2651	1.6673	45
50	281.03	0.01727	8.52	250.08	1095.6	250.24	924.2	1174.4	0.4113	1.2476	1.6589	50
55	287.10	0.01733	7.79	256.28	1097.0	256.46	919.9	1176.3	0.4196	1.2317	1.6513	55
60	292.73	0.01738	7.177	262.1	1098.3	262.2	915.8	1178.0	0.4273	1.2170	1.6443	60
65	298.00	0.01743	6.647	267.5	1099.5	267.7	911.9	1179.6	0.4345	1.2035	1.6380	65
70	302.96	0.01748	6.209	272.6	1100.6	272.8	908.3	1181.0	0.4412	1.1909	1.6321	70
75	307.63	0.01752	5.818	277.4	1101.6	277.6	904.8	1182.4	0.4475	1.1790	1.6265	75
80	312.07	0.01757	5.474	282.0	1102.6	282.2	901.4	1183.6	0.4534	1.1679	1.6213	80
85	316.29	0.01761	5.170	286.3	1103.5	286.6	898.2	1184.8	0.4591	1.1574	1.6165	85
90	320.31	0.01766	4.898	290.5	1104.3	290.8	895.1	1185.9	0.4644	1.1475	1.6119	90
95	324.16	0.01770	4.654	294.5	1105.0	294.8	892.1	1186.9	0.4695	1.1380	1.6075	95
100	327.86	0.01774	4.434	298.3	1105.8	298.6	889.2	1187.8	0.4744	1.1290	1.6034	100
110	334.82	0.01781	4.051	305.5	1107.1	305.9	883.7	1189.6	0.4836	1.1122	1.5958	110
120	341.30	0.01789	3.730	312.3	1108.3	312.7	878.5	1191.1	0.4920	1.0966	1.5886	120
130	347.37	0.01796	3.457	318.6	1109.4	319.0	873.5	1192.5	0.4999	1.0822	1.5821	130
140	353.08	0.01802	3.221	324.6	1110.3	325.1	868.7	1193.8	0.5073	1.0688	1.5761	140
150	358.48	0.01809	3.016	330.2	1111.2	330.8	864.2	1194.9	0.5142	1.0562	1.5704	150
160	363.60	0.01815	2.836	335.6	1112.0	336.2	859.8	1196.0	0.5208	1.0443	1.5651	160

H$_2$O

TABLE A-3E

(Continued)

Press. lbf/in.^{2}	Temp. °F	Specific Volume ft^3/lb		Internal Energy Btu/lb		Enthalpy Btu/lb			Entropy Btu/lb·°R			Press. lbf/in.^{2}
		Sat. Liquid v_f	Sat. Vapor v_g	Sat. Liquid u_f	Sat. Vapor u_g	Sat. Liquid h_f	Evap. h_{fg}	Sat. Vapor h_g	Sat. Liquid s_f	Evap. s_{fg}	Sat. Vapor s_g	
170	368.47	0.01821	2.676	340.8	1112.7	341.3	855.6	1196.9	0.5270	1.0330	1.5600	170
180	373.13	0.01827	2.553	345.7	1113.4	346.3	851.5	1197.8	0.5329	1.0223	1.5552	180
190	377.59	0.01833	2.405	350.4	1114.0	351.0	847.5	1198.6	0.5386	1.0122	1.5508	190
200	381.86	0.01839	2.289	354.9	1114.6	355.6	843.7	1199.3	0.5440	1.0025	1.5465	200
250	401.04	0.01865	1.845	375.4	1116.7	376.2	825.8	1202.1	0.5680	0.9594	1.5274	250
300	417.43	0.01890	1.544	393.0	1118.2	394.1	809.8	1203.9	0.5883	0.9232	1.5115	300
350	431.82	0.01912	1.327	408.7	1119.0	409.9	795.0	1204.9	0.6060	0.8917	1.4977	350
400	444.70	0.01934	1.162	422.8	1119.5	424.2	781.2	1205.5	0.6218	0.8638	1.4856	400
450	456.39	0.01955	1.033	435.7	1119.6	437.4	768.2	1205.6	0.6360	0.8385	1.4745	450
500	467.13	0.01975	0.928	447.7	1119.4	449.5	755.8	1205.3	0.6490	0.8154	1.4644	500
550	477.07	0.01994	0.842	458.9	1119.1	460.9	743.9	1204.8	0.6611	0.7941	1.4451	550
600	486.33	0.02013	0.770	469.4	1118.6	471.7	732.4	1204.1	0.6723	0.7742	1.4464	600
700	503.23	0.02051	0.656	488.9	1117.0	491.5	710.5	1202.0	0.6927	0.7378	1.4305	700
800	518.36	0.02087	0.569	506.6	1115.0	509.7	689.6	1199.3	0.7110	0.7050	1.4160	800
900	532.12	0.02123	0.501	523.0	1112.6	526.6	669.5	1196.0	0.7277	0.6750	1.4027	900
1000	544.75	0.02159	0.446	538.4	1109.9	542.4	650.0	1192.4	0.7432	0.6471	1.3903	1000
1100	556.45	0.02195	0.401	552.9	1106.8	557.4	631.0	1188.3	0.7576	0.6209	1.3786	1100
1200	567.37	0.02232	0.362	566.7	1103.5	571.7	612.3	1183.9	0.7712	0.5961	1.3673	1200
1300	577.60	0.02269	0.330	579.9	1099.8	585.4	593.8	1179.2	0.7841	0.5724	1.3565	1300
1400	587.25	0.02307	0.302	592.7	1096.0	598.6	575.5	1174.1	0.7964	0.5497	1.3461	1400
1500	596.39	0.02346	0.277	605.0	1091.8	611.5	557.2	1168.7	0.8082	0.5276	1.3359	1500
1600	605.06	0.02386	0.255	616.9	1087.4	624.0	538.9	1162.9	0.8196	0.5062	1.3258	1600
1700	613.32	0.02428	0.236	628.6	1082.7	636.2	520.6	1156.9	0.8307	0.4852	1.3159	1700
1800	621.21	0.02472	0.218	640.0	1077.7	648.3	502.1	1150.4	0.8414	0.4645	1.3060	1800
1900	628.76	0.02517	0.203	651.3	1072.3	660.1	483.4	1143.5	0.8519	0.4441	1.2961	1900
2000	636.00	0.02565	0.188	662.4	1066.6	671.9	464.4	1136.3	0.8623	0.4238	1.2861	2000
2250	652.90	0.02698	0.157	689.9	1050.6	701.1	414.8	1115.9	0.8876	0.3728	1.2604	2250
2500	668.31	0.02860	0.131	717.7	1031.0	730.9	360.5	1091.4	0.9131	0.3196	1.2327	2500
2750	682.46	0.03077	0.107	747.3	1005.9	763.0	297.4	1060.4	0.9401	0.2604	1.2005	2750
3000	695.52	0.03431	0.084	783.4	968.8	802.5	213.0	1015.5	0.9732	0.1843	1.1575	3000
3203.6	705.44	0.05053	0.0505	872.6	872.6	902.5	0	902.5	1.0580	0	1.0580	3203.6

TABLE A-4E

Properties of Superheated Water Vapor

H₂O

T °F	v ft³/lb	u Btu/lb	h Btu/lb	s Btu/lb·°R	v ft³/lb	u Btu/lb	h Btu/lb	s Btu/lb·°R

T °F	v ft³/lb	u Btu/lb	h Btu/lb	s Btu/lb·°R	v ft³/lb	u Btu/lb	h Btu/lb	s Btu/lb·°R
	$p = 1$ lbf/in.² ($T_{sat} = 101.7$°F)				$p = 5$ lbf/in.² ($T_{sat} = 162.2$°F)			
Sat.	333.6	1044.0	1105.8	1.9779	73.53	1063.0	1131.0	1.8441
150	362.6	1060.4	1127.5	2.0151				
200	392.5	1077.5	1150.1	2.0508	78.15	1076.0	1148.6	1.8715
250	422.4	1094.7	1172.8	2.0839	84.21	1093.8	1171.7	1.9052
300	452.3	1112.0	1195.7	2.1150	90.24	1111.3	1194.8	1.9367
400	511.9	1147.0	1241.8	2.1720	102.24	1146.6	1241.2	1.9941
500	571.5	1182.8	1288.5	2.2235	114.20	1182.5	1288.2	2.0458
600	631.1	1219.3	1336.1	2.2706	126.15	1219.1	1335.8	2.0930
700	690.7	1256.7	1384.5	2.3142	138.08	1256.5	1384.3	2.1367
800	750.3	1294.4	1433.7	2.3550	150.01	1294.7	1433.5	2.1775
900	809.9	1333.9	1483.8	2.3932	161.94	1333.8	1483.7	2.2158
1000	869.5	1373.9	1534.8	2.4294	173.86	1373.9	1534.7	2.2520
	$p = 10$ lbf/in.² ($T_{sat} = 193.2$°F)				$p = 14.7$ lbf/in.² ($T_{sat} = 212.0$°F)			
Sat.	38.42	1072.2	1143.3	1.7877	26.80	1077.6	1150.5	1.7567
200	38.85	1074.7	1146.6	1.7927				
250	41.95	1092.6	1170.2	1.8272	28.42	1091.5	1168.8	1.7832
300	44.99	1110.4	1193.7	1.8592	30.52	1109.6	1192.6	1.8157
400	51.03	1146.1	1240.5	1.9171	34.67	1145.6	1239.9	1.8741
500	57.04	1182.2	1287.7	1.9690	38.77	1181.8	1287.3	1.9263
600	63.03	1218.9	1335.5	2.0164	42.86	1218.6	1335.2	1.9737
700	69.01	1256.3	1384.0	2.0601	46.93	1256.1	1383.8	2.0175
800	74.98	1294.6	1433.3	2.1009	51.00	1294.4	1433.1	2.0584
900	80.95	1333.7	1483.5	2.1393	55.07	1333.6	1483.4	2.0967
1000	86.91	1373.8	1534.6	2.1755	59.13	1373.7	1534.5	2.1330
1100	92.88	1414.7	1586.6	2.2099	63.19	1414.6	1586.4	2.1674
	$p = 20$ lbf/in.² ($T_{sat} = 228.0$°F)				$p = 40$ lbf/in.² ($T_{sat} = 267.3$°F)			
Sat.	20.09	1082.0	1156.4	1.7320	10.50	1093.3	1170.0	1.6767
250	20.79	1090.3	1167.2	1.7475				
300	22.36	1108.7	1191.5	1.7805	11.04	1105.1	1186.8	1.6993
350	23.90	1126.9	1215.4	1.8110	11.84	1124.2	1211.8	1.7312
400	25.43	1145.1	1239.2	1.8395	12.62	1143.0	1236.4	1.7606
500	28.46	1181.5	1286.8	1.8919	14.16	1180.1	1284.9	1.8140
600	31.47	1218.4	1334.8	1.9395	15.69	1217.3	1333.4	1.8621
700	34.47	1255.9	1383.5	1.9834	17.20	1255.1	1382.4	1.9063
800	37.46	1294.3	1432.9	2.0243	18.70	1293.7	1432.1	1.9474
900	40.45	1333.5	1483.2	2.0627	20.20	1333.0	1482.5	1.9859
1000	43.44	1373.5	1534.3	2.0989	21.70	1373.1	1533.8	2.0223
1100	46.42	1414.5	1586.3	2.1334	23.20	1414.2	1585.9	2.0568

H₂O

TABLE A-4E

(Continued)

T °F	v ft³/lb	u Btu/lb	h Btu/lb	s Btu/lb·°R	v ft³/lb	u Btu/lb	h Btu/lb	s Btu/lb·°R
		$p = 60$ lbf/in.² ($T_{sat} = 292.7$°F)				$p = 80$ lbf/in.² ($T_{sat} = 312.1$°F)		
Sat.	7.17	1098.3	1178.0	1.6444	5.47	1102.6	1183.6	1.6214
300	7.26	1101.3	1181.9	1.6496				
350	7.82	1121.4	1208.2	1.6830	5.80	1118.5	1204.3	1.6476
400	8.35	1140.8	1233.5	1.7134	6.22	1138.5	1230.6	1.6790
500	9.40	1178.6	1283.0	1.7678	7.02	1177.2	1281.1	1.7346
600	10.43	1216.3	1332.1	1.8165	7.79	1215.3	1330.7	1.7838
700	11.44	1254.4	1381.4	1.8609	8.56	1253.6	1380.3	1.8285
800	12.45	1293.0	1431.2	1.9022	9.32	1292.4	1430.4	1.8700
900	13.45	1332.5	1481.8	1.9408	10.08	1332.0	1481.2	1.9087
1000	14.45	1372.7	1533.2	1.9773	10.83	1372.3	1532.6	1.9453
1100	15.45	1413.8	1585.4	2.0119	11.58	1413.5	1584.9	1.9799
1200	16.45	1455.8	1638.5	2.0448	12.33	1455.5	1638.1	2.0130
		$p = 100$ lbf/in.² ($T_{sat} = 327.8$°F)				$p = 120$ lbf/in.² ($T_{sat} = 341.3$°F)		
Sat.	4.434	1105.8	1187.8	1.6034	3.730	1108.3	1191.1	1.5886
350	4.592	1115.4	1200.4	1.6191	3.783	1112.2	1196.2	1.5950
400	4.934	1136.2	1227.5	1.6517	4.079	1133.8	1224.4	1.6288
450	5.265	1156.2	1253.6	1.6812	4.360	1154.3	1251.2	1.6590
500	5.587	1175.7	1279.1	1.7085	4.633	1174.2	1277.1	1.6868
600	6.216	1214.2	1329.3	1.7582	5.164	1213.2	1327.8	1.7371
700	6.834	1252.8	1379.2	1.8033	5.682	1252.0	1378.2	1.7825
800	7.445	1291.8	1429.6	1.8449	6.195	1291.2	1428.7	1.8243
900	8.053	1331.5	1480.5	1.8838	6.703	1330.9	1479.8	1.8633
1000	8.657	1371.9	1532.1	1.9204	7.208	1371.5	1531.5	1.9000
1100	9.260	1413.1	1584.5	1.9551	7.711	1412.8	1584.0	1.9348
1200	9.861	1455.2	1637.7	1.9882	8.213	1454.9	1637.3	1.9679
		$p = 140$ lbf/in.² ($T_{sat} = 353.1$°F)				$p = 160$ lbf/in.² ($T_{sat} = 363.6$°F)		
Sat.	3.221	1110.3	1193.8	1.5761	2.836	1112.0	1196.0	1.5651
400	3.466	1131.4	1221.2	1.6088	3.007	1128.8	1217.8	1.5911
450	3.713	1152.4	1248.6	1.6399	3.228	1150.5	1246.1	1.6230
500	3.952	1172.7	1275.1	1.6682	3.440	1171.2	1273.0	1.6518
550	4.184	1192.5	1300.9	1.6945	3.646	1191.3	1299.2	1.6785
600	4.412	1212.1	1326.4	1.7191	3.848	1211.1	1325.0	1.7034
700	4.860	1251.2	1377.1	1.7648	4.243	1250.4	1376.0	1.7494
800	5.301	1290.5	1427.9	1.8068	4.631	1289.9	1427.0	1.7916
900	5.739	1330.4	1479.1	1.8459	5.015	1329.9	1478.4	1.8308
1000	6.173	1371.0	1531.0	1.8827	5.397	1370.6	1530.4	1.8677
1100	6.605	1412.4	1583.6	1.9176	5.776	1412.1	1583.1	1.9026
1200	7.036	1454.6	1636.9	1.9507	6.154	1454.3	1636.5	1.9358

TABLE A-4E

(*Continued*)

T °F	v ft³/lb	u Btu/lb	h Btu/lb	s Btu/lb·°R	v ft³/lb	u Btu/lb	h Btu/lb	s Btu/lb·°R
	p = 180 lbf/in.² (T_sat = 373.1°F)				p = 200 lbf/in.² (T_sat = 381.8°F)			
Sat.	2.533	1113.4	1197.8	1.5553	2.289	1114.6	1199.3	1.5464
400	2.648	1126.2	1214.4	1.5749	2.361	1123.5	1210.8	1.5600
450	2.850	1148.5	1243.4	1.6078	2.548	1146.4	1240.7	1.5938
500	3.042	1169.6	1270.9	1.6372	2.724	1168.0	1268.8	1.6239
550	3.228	1190.0	1297.5	1.6642	2.893	1188.7	1295.7	1.6512
600	3.409	1210.0	1323.5	1.6893	3.058	1208.9	1322.1	1.6767
700	3.763	1249.6	1374.9	1.7357	3.379	1248.8	1373.8	1.7234
800	4.110	1289.3	1426.2	1.7781	3.693	1288.6	1425.3	1.7660
900	4.453	1329.4	1477.7	1.8174	4.003	1328.9	1477.1	1.8055
1000	4.793	1370.2	1529.8	1.8545	4.310	1369.8	1529.3	1.8425
1100	5.131	1411.7	1582.6	1.8894	4.615	1411.4	1582.2	1.8776
1200	5.467	1454.0	1636.1	1.9227	4.918	1453.7	1635.7	1.9109
	p = 250 lbf/in.² (T_sat = 401.0°F)				p = 300 lbf/in.² (T_sat = 417.4°F)			
Sat.	1.845	1116.7	1202.1	1.5274	1.544	1118.2	1203.9	1.5115
450	2.002	1141.1	1233.7	1.5632	1.636	1135.4	1226.2	1.5365
500	2.150	1163.8	1263.3	1.5948	1.766	1159.5	1257.5	1.5701
550	2.290	1185.3	1291.3	1.6233	1.888	1181.9	1286.7	1.5997
600	2.426	1206.1	1318.3	1.6494	2.004	1203.2	1314.5	1.6266
700	2.688	1246.7	1371.1	1.6970	2.227	1244.0	1368.3	1.6751
800	2.943	1287.0	1423.2	1.7301	2.442	1285.4	1421.0	1.7187
900	3.193	1327.6	1475.3	1.7799	2.653	1326.3	1473.6	1.7589
1000	3.440	1368.7	1527.9	1.8172	2.860	1367.7	1526.5	1.7964
1100	3.685	1410.5	1581.0	1.8524	3.066	1409.6	1579.8	1.8317
1200	3.929	1453.0	1634.8	1.8858	3.270	1452.2	1633.8	1.8653
1300	4.172	1496.3	1689.3	1.9177	3.473	1495.6	1688.4	1.8973
	p = 350 lbf/in.² (T_sat = 431.8°F)				p = 400 lbf/in.² (T_sat = 444.7°F)			
Sat.	1.327	1119.0	1204.9	1.4978	1.162	1119.5	1205.5	1.4856
450	1.373	1129.2	1218.2	1.5125	1.175	1122.6	1209.5	1.4901
500	1.491	1154.9	1251.5	1.5482	1.284	1150.1	1245.2	1.5282
550	1.600	1178.3	1281.9	1.5790	1.383	1174.6	1277.0	1.5605
600	1.703	1200.3	1310.6	1.6068	1.476	1197.3	1306.6	1.5892
700	1.898	1242.5	1365.4	1.6562	1.650	1240.4	1362.5	1.6397
800	2.085	1283.8	1418.8	1.7004	1.816	1282.1	1416.6	1.6844
900	2.267	1325.0	1471.8	1.7409	1.978	1323.7	1470.1	1.7252
1000	2.446	1366.6	1525.0	1.7787	2.136	1365.5	1523.6	1.7632
1100	2.624	1408.7	1578.6	1.8142	2.292	1407.8	1577.4	1.7989
1200	2.799	1451.5	1632.8	1.8478	2.446	1450.7	1621.8	1.8327
1300	2.974	1495.0	1687.6	1.8799	2.599	1494.3	1686.8	1.8648

H₂O

TABLE A-4E

(Continued)

T °F	v ft³/lb	u Btu/lb	h Btu/lb	s Btu/lb·°R	v ft³/lb	u Btu/lb	h Btu/lb	s Btu/lb·°R
	$p = 450$ lbf/in.² ($T_{sat} = 456.4°F$)				$p = 500$ lbf/in.² ($T_{sat} = 467.1°F$)			
Sat.	1.033	1119.6	1205.6	1.4746	0.928	1119.4	1205.3	1.4645
500	1.123	1145.1	1238.5	1.5097	0.992	1139.7	1231.5	1.4923
550	1.215	1170.7	1271.9	1.5436	1.079	1166.7	1266.6	1.5279
600	1.300	1194.3	1302.5	1.5732	1.158	1191.1	1298.3	1.5585
700	1.458	1238.2	1359.6	1.6248	1.304	1236.0	1356.7	1.6112
800	1.608	1280.5	1414.4	1.6701	1.441	1278.8	1412.1	1.6571
900	1.752	1322.4	1468.3	1.7113	1.572	1321.0	1466.5	1.6987
1000	1.894	1364.4	1522.2	1.7495	1.701	1363.3	1520.7	1.7371
1100	2.034	1406.9	1576.3	1.7853	1.827	1406.0	1575.1	1.7731
1200	2.172	1450.0	1630.8	1.8192	1.952	1449.2	1629.8	1.8072
1300	2.308	1493.7	1685.9	1.8515	2.075	1493.1	1685.1	1.8395
1400	2.444	1538.1	1741.7	1.8823	2.198	1537.6	1741.0	1.8704
	$p = 600$ lbf/in.² ($T_{sat} = 486.3°F$)				$p = 700$ lbf/in.² ($T_{sat} = 503.2°F$)			
Sat.	0.770	1118.6	1204.1	1.4464	0.656	1117.0	1202.0	1.4305
500	0.795	1128.0	1216.2	1.4592				
550	0.875	1158.2	1255.4	1.4990	0.728	1149.0	1243.2	1.4723
600	0.946	1184.5	1289.5	1.5320	0.793	1177.5	1280.2	1.5081
700	1.073	1231.5	1350.6	1.5872	0.907	1226.9	1344.4	1.5661
800	1.190	1275.4	1407.6	1.6343	1.011	1272.0	1402.9	1.6145
900	1.302	1318.4	1462.9	1.6766	1.109	1315.6	1459.3	1.6576
1000	1.411	1361.2	1517.8	1.7155	1.204	1358.9	1514.9	1.6970
1100	1.517	1404.2	1572.7	1.7519	1.296	1402.4	1570.2	1.7337
1200	1.622	1447.7	1627.8	1.7861	1.387	1446.2	1625.8	1.7682
1300	1.726	1491.7	1683.4	1.8186	1.476	1490.4	1681.7	1.8009
1400	1.829	1536.5	1739.5	1.8497	1.565	1535.3	1738.1	1.8321
	$p = 800$ lbf/in.² ($T_{sat} = 518.3°F$)				$p = 900$ lbf/in.² ($T_{sat} = 532.1°F$)			
Sat.	0.569	1115.0	1199.3	1.4160	0.501	1112.6	1196.0	1.4027
550	0.615	1138.8	1229.9	1.4469	0.527	1127.5	1215.2	1.4219
600	0.677	1170.1	1270.4	1.4861	0.587	1162.2	1260.0	1.4652
650	0.732	1197.2	1305.6	1.5186	0.639	1191.1	1297.5	1.4999
700	0.783	1222.1	1338.0	1.5471	0.686	1217.1	1331.4	1.5297
800	0.876	1268.5	1398.2	1.5969	0.772	1264.9	1393.4	1.5810
900	0.964	1312.9	1455.6	1.6408	0.851	1310.1	1451.9	1.6257
1000	1.048	1356.7	1511.9	1.6807	0.927	1354.5	1508.9	1.6662
1100	1.130	1400.5	1567.8	1.7178	1.001	1398.7	1565.4	1.7036
1200	1.210	1444.6	1623.8	1.7526	1.073	1443.0	1621.7	1.7386
1300	1.289	1489.1	1680.0	1.7854	1.144	1487.8	1687.3	1.7717
1400	1.367	1534.2	1736.6	1.8167	1.214	1533.0	1735.1	1.8031

TABLE A-4E

(Continued)

T °F	v ft³/lb	u Btu/lb	h Btu/lb	s Btu/lb · °R	v ft³/lb	u Btu/lb	h Btu/lb	s Btu/lb · °R
	$p = 1000$ lbf/in.² ($T_{sat} = 544.7°F$)				$p = 1200$ lbf/in.² ($T_{sat} = 567.4°F$)			
Sat.	0.446	1109.0	1192.4	1.3903	0.362	1103.5	1183.9	1.3673
600	0.514	1153.7	1248.8	1.4450	0.402	1134.4	1223.6	1.4054
650	0.564	1184.7	1289.1	1.4822	0.450	1170.9	1270.8	1.4490
700	0.608	1212.0	1324.6	1.5135	0.491	1201.3	1310.2	1.4837
800	0.688	1261.2	1388.5	1.5665	0.562	1253.7	1378.4	1.5402
900	0.761	1307.3	1448.1	1.6120	0.626	1301.5	1440.4	1.5876
1000	0.831	1352.2	1505.9	1.6530	0.685	1347.5	1499.7	1.6297
1100	0.898	1396.8	1562.9	1.6908	0.743	1393.0	1557.9	1.6682
1200	0.963	1441.5	1619.7	1.7261	0.798	1438.3	1615.5	1.7040
1300	1.027	1486.5	1676.5	1.7593	0.853	1483.8	1673.1	1.7377
1400	1.091	1531.9	1733.7	1.7909	0.906	1529.6	1730.7	1.7696
1600	1.215	1624.4	1849.3	1.8499	1.011	1622.6	1847.1	1.8290
	$p = 1400$ lbf/in.² ($T_{sat} = 587.2°F$)				$p = 1600$ lbf/in.² ($T_{sat} = 605.1°F$)			
Sat.	0.302	1096.0	1174.1	1.3461	0.255	1087.4	1162.9	1.3258
600	0.318	1110.9	1193.1	1.3641				
650	0.367	1155.5	1250.5	1.4171	0.303	1137.8	1227.4	1.3852
700	0.406	1189.6	1294.8	1.4562	0.342	1177.0	1278.1	1.4299
800	0.471	1245.8	1367.9	1.5168	0.403	1237.7	1357.0	1.4953
900	0.529	1295.6	1432.5	1.5661	0.466	1289.5	1424.4	1.5468
1000	0.582	1342.8	1493.5	1.6094	0.504	1338.0	1487.1	1.5913
1100	0.632	1389.1	1552.8	1.6487	0.549	1385.2	1547.7	1.6315
1200	0.681	1435.1	1611.4	1.6851	0.592	1431.8	1607.1	1.6684
1300	0.728	1481.1	1669.6	1.7192	0.634	1478.3	1666.1	1.7029
1400	0.774	1527.2	1727.8	1.7513	0.675	1524.9	1724.8	1.7354
1600	0.865	1620.8	1844.8	1.8111	0.755	1619.0	1842.6	1.7955
	$p = 1800$ lbf/in.² ($T_{sat} = 621.2°F$)				$p = 2000$ lbf/in.² ($T_{sat} = 636.0°F$)			
Sat.	0.218	1077.7	1150.4	1.3060	0.188	1066.6	1136.3	1.2861
650	0.251	1117.0	1200.4	1.3517	0.206	1091.1	1167.2	1.3141
700	0.291	1163.1	1259.9	1.4042	0.249	1147.7	1239.8	1.3782
750	0.322	1198.6	1305.9	1.4430	0.280	1187.3	1291.1	1.4216
800	0.350	1229.1	1345.7	1.4753	0.307	1220.1	1333.8	1.4562
900	0.399	1283.2	1416.1	1.5291	0.353	1276.8	1407.6	1.5126
1000	0.443	1333.1	1480.7	1.5749	0.395	1328.1	1474.1	1.5598
1100	0.484	1381.2	1542.5	1.6159	0.433	1377.2	1537.2	1.6017
1200	0.524	1428.5	1602.9	1.6534	0.469	1425.2	1598.6	1.6398
1300	0.561	1475.5	1662.5	1.6883	0.503	1472.7	1659.0	1.6751
1400	0.598	1522.5	1721.8	1.7211	0.537	1520.2	1718.8	1.7082
1600	0.670	1617.2	1840.4	1.7817	0.602	1615.4	1838.2	1.7692

H₂O

TABLE A-4E

(*Continued*)

T °F	v ft³/lb	u Btu/lb	h Btu/lb	s Btu/lb · °R	v ft³/lb	u Btu/lb	h Btu/lb	s Btu/lb · °R
	p = 2500 lbf/in.² (T_sat = 668.3°F)				p = 3000 lbf/in.² (T_sat = 695.5°F)			
Sat.	0.1306	1031.0	1091.4	1.2327	0.0840	968.8	1015.5	1.1575
700	0.1684	1098.7	1176.6	1.3073	0.0977	1003.9	1058.1	1.1944
750	0.2030	1155.2	1249.1	1.3686	0.1483	1114.7	1197.1	1.3122
800	0.2291	1195.7	1301.7	1.4112	0.1757	1167.6	1265.2	1.3675
900	0.2712	1259.9	1385.4	1.4752	0.2160	1241.8	1361.7	1.4414
1000	0.3069	1315.2	1457.2	1.5262	0.2485	1301.7	1439.6	1.4967
1100	0.3393	1366.8	1523.8	1.5704	0.2772	1356.2	1510.1	1.5434
1200	0.3696	1416.7	1587.7	1.6101	0.3086	1408.0	1576.6	1.5848
1300	0.3984	1465.7	1650.0	1.6465	0.3285	1458.5	1640.9	1.6224
1400	0.4261	1514.2	1711.3	1.6804	0.3524	1508.1	1703.7	1.6571
1500	0.4531	1562.5	1772.1	1.7123	0.3754	1557.3	1765.7	1.6896
1600	0.4795	1610.8	1832.6	1.7424	0.3978	1606.3	1827.1	1.7201
	p = 3500 lbf/in.²				p = 4000 lbf/in.²			
650	0.0249	663.5	679.7	0.8630	0.0245	657.7	675.8	0.8574
700	0.0306	759.5	779.3	0.9506	0.0287	742.1	763.4	0.9345
750	0.1046	1058.4	1126.1	1.2440	0.0633	960.7	1007.5	1.1395
800	0.1363	1134.7	1223.0	1.3226	0.1052	1095.0	1172.9	1.2740
900	0.1763	1222.4	1336.5	1.4096	0.1462	1201.5	1309.7	1.3789
1000	0.2066	1287.6	1421.4	1.4699	0.1752	1272.9	1402.6	1.4449
1100	0.2328	1345.2	1496.0	1.5193	0.1995	1333.9	1481.6	1.4973
1200	0.2566	1399.2	1565.3	1.5624	0.2213	1390.1	1553.9	1.5423
1300	0.2787	1451.1	1631.7	1.6012	0.2414	1443.7	1622.4	1.5823
1400	0.2997	1501.9	1696.1	1.6368	0.2603	1495.7	1688.4	1.6188
1500	0.3199	1552.0	1759.2	1.6699	0.2784	1546.7	1752.8	1.6526
1600	0.3395	1601.7	1831.6	1.7010	0.2959	1597.1	1816.1	1.6841
	p = 4400 lbf/in.²				p = 4800 lbf/in.²			
650	0.0242	653.6	673.3	0.8535	0.0237	649.8	671.0	0.8499
700	0.0278	732.7	755.3	0.9257	0.0271	725.1	749.1	0.9187
750	0.0415	870.8	904.6	1.0513	0.0352	832.6	863.9	1.0154
800	0.0844	1056.5	1125.3	1.2306	0.0668	1011.2	1070.5	1.1827
900	0.1270	1183.7	1287.1	1.3548	0.1109	1164.8	1263.4	1.3310
1000	0.1552	1260.8	1387.2	1.4260	0.1385	1248.3	1317.4	1.4078
1100	0.1784	1324.7	1469.9	1.4809	0.1608	1315.3	1458.1	1.4653
1200	0.1989	1382.8	1544.7	1.5274	0.1802	1375.4	1535.4	1.5133
1300	0.2176	1437.7	1614.9	1.5685	0.1979	1431.7	1607.4	1.5555
1400	0.2352	1490.7	1682.3	1.6057	0.2143	1485.7	1676.1	1.5934
1500	0.2520	1542.7	1747.6	1.6399	0.2300	1538.2	1742.5	1.6282
1600	0.2681	1593.4	1811.7	1.6718	0.2450	1589.8	1807.4	1.6605

TABLE A-5E

Properties of Compressed Liquid Water

T °F	v ft³/lb	u Btu/lb	h Btu/lb	s Btu/lb·°R	v ft³/lb	u Btu/lb	h Btu/lb	s Btu/lb·°R
	$p = 500$ lbf/in.² ($T_{sat} = 467.1°F$)				$p = 1000$ lbf/in.² ($T_{sat} = 544.7°F$)			
32	0.015994	0.00	1.49	0.00000	0.015967	0.03	2.99	0.00005
50	0.015998	18.02	19.50	0.03599	0.015972	17.99	20.94	0.03592
100	0.016106	67.87	69.36	0.12932	0.016082	67.70	70.68	0.12901
150	0.016318	117.66	119.17	0.21457	0.016293	117.38	120.40	0.21410
200	0.016608	167.65	169.19	0.29341	0.016580	167.26	170.32	0.29281
300	0.017416	268.92	270.53	0.43641	0.017379	268.24	271.46	0.43552
400	0.018608	373.68	375.40	0.56604	0.018550	372.55	375.98	0.56472
Sat.	0.019748	447.70	449.53	0.64904	0.021591	538.39	542.38	0.74320

T °F	v ft³/lb	u Btu/lb	h Btu/lb	s Btu/lb·°R	v ft³/lb	u Btu/lb	h Btu/lb	s Btu/lb·°R
	$p = 1500$ lbf/in.² ($T_{sat} = 596.4°F$)				$p = 2000$ lbf/in.² ($T_{sat} = 636.0°F$)			
32	0.015939	0.05	4.47	0.00007	0.015912	0.06	5.95	0.00008
50	0.015946	17.95	22.38	0.03584	0.015920	17.91	23.81	0.03575
100	0.016058	67.53	71.99	0.12870	0.016034	67.37	73.30	0.12839
150	0.016268	117.10	121.62	0.21364	0.016244	116.83	122.84	0.21318
200	0.016554	166.87	171.46	0.29221	0.016527	166.49	172.60	0.29162
300	0.017343	267.58	272.39	0.43463	0.017308	266.93	273.33	0.43376
400	0.018493	371.45	376.59	0.56343	0.018439	370.38	377.21	0.56216
500	0.02024	481.8	487.4	0.6853	0.02014	479.8	487.3	0.6832
Sat.	0.02346	605.0	611.5	0.8082	0.02565	662.4	671.9	0.8623

T °F	v ft³/lb	u Btu/lb	h Btu/lb	s Btu/lb·°R	v ft³/lb	u Btu/lb	h Btu/lb	s Btu/lb·°R
	$p = 3000$ lbf/in.² ($T_{sat} = 695.5°F$)				$p = 4000$ lbf/in.²			
32	0.015859	0.09	8.90	0.00009	0.015807	0.10	11.80	0.00005
50	0.015870	17.84	26.65	0.03555	0.015821	17.76	29.47	0.03534
100	0.015987	67.04	75.91	0.12777	0.015942	66.72	78.52	0.12714
150	0.016196	116.30	125.29	0.21226	0.016150	115.77	127.73	0.21136
200	0.016476	165.74	174.89	0.29046	0.016425	165.02	177.18	0.28931
300	0.017240	265.66	275.23	0.43205	0.017174	264.43	277.15	0.43038
400	0.018334	368.32	378.50	0.55970	0.018235	366.35	379.85	0.55734
500	0.019944	476.2	487.3	0.6794	0.019766	472.9	487.5	0.6758
Sat.	0.034310	783.5	802.5	0.9732				

H₂O

H₂O

TABLE A-6E

Properties of Saturated Water (Solid–Vapor): Temperature Table

Temp. °F	Press. lbf/in.²	Specific Volume ft³/lb		Internal Energy Btu/lb			Enthalpy Btu/lb			Entropy Btu/lb · °R		
		Sat. Solid v_i	Sat. Vapor $v_g \times 10^{-3}$	Sat. Solid u_i	Subl. u_{ig}	Sat. Vapor u_g	Sat. Solid h_i	Subl. h_{ig}	Sat. Vapor h_g	Sat. Solid s_i	Subl. s_{ig}	Sat. Vapor s_g
32.018	.0887	.01747	3.302	−143.34	1164.6	1021.2	−143.34	1218.7	1075.4	−.292	2.479	2.187
32	.0886	.01747	3.305	−143.35	1164.6	1021.2	−143.35	1218.7	1075.4	−.292	2.479	2.187
30	.0808	.01747	3.607	−144.35	1164.9	1020.5	−144.35	1218.9	1074.5	−.294	2.489	2.195
25	.0641	.01746	4.506	−146.84	1165.7	1018.9	−146.84	1219.1	1072.3	−.299	2.515	2.216
20	.0505	.01745	5.655	−149.31	1166.5	1017.2	−149.31	1219.4	1070.1	−.304	2.542	2.238
15	.0396	.01745	7.13	−151.75	1167.3	1015.5	−151.75	1219.7	1067.9	−.309	2.569	2.260
10	.0309	.01744	9.04	−154.17	1168.1	1013.9	−154.17	1219.9	1065.7	−.314	2.597	2.283
5	.0240	.01743	11.52	−156.56	1168.8	1012.2	−156.56	1220.1	1063.5	−.320	2.626	2.306
0	.0185	.01743	14.77	−158.93	1169.5	1010.6	−158.93	1220.2	1061.2	−.325	2.655	2.330
−5	.0142	.01742	19.03	−161.27	1170.2	1008.9	−161.27	1220.3	1059.0	−.330	2.684	2.354
−10	.0109	.01741	24.66	−163.59	1170.9	1007.3	−163.59	1220.4	1056.8	−.335	2.714	2.379
−15	.0082	.01740	32.2	−165.89	1171.5	1005.6	−165.89	1220.5	1054.6	−.340	2.745	2.405
−20	.0062	.01740	42.2	−168.16	1172.1	1003.9	−168.16	1220.6	1052.4	−.345	2.776	2.431
−25	.0046	.01739	55.7	−170.40	1172.7	1002.3	−170.40	1220.6	1050.2	−.351	2.808	2.457
−30	.0035	.01738	74.1	−172.63	1173.2	1000.6	−172.63	1220.6	1048.0	−.356	2.841	2.485
−35	.0026	.01737	99.2	−174.82	1173.8	998.9	−174.82	1220.6	1045.8	−.361	2.874	2.513
−40	.0019	.01737	133.8	−177.00	1174.3	997.3	−177.00	1220.6	1043.6	−.366	2.908	2.542

TABLE A-7E

Properties of Saturated Refrigerant 22 (Liquid–Vapor): Temperature Table

Temp. °F	Press. lbf/in.^{2}	Specific Volume ft^3/lb Sat. Liquid v_f	Sat. Vapor v_g	Internal Energy Btu/lb Sat. Liquid u_f	Sat. Vapor u_g	Enthalpy Btu/lb Sat. Liquid h_f	Evap. h_{fg}	Sat. Vapor h_g	Entropy Btu/lb · °R Sat. Liquid s_f	Sat. Vapor s_g	Temp. °F
−80	4.781	0.01090	9.6984	−10.30	87.24	−10.29	106.11	95.82	−0.0257	0.2538	−80
−60	8.834	0.01113	5.4744	−5.20	89.16	−5.18	103.30	98.12	−0.0126	0.2458	−60
−55	10.187	0.01120	4.7933	−3.91	89.64	−3.89	102.58	98.68	−0.0094	0.2441	−55
−50	11.701	0.01126	4.2123	−2.62	90.12	−2.60	101.84	99.24	−0.0063	0.2424	−50
−45	13.387	0.01132	3.7147	−1.33	90.59	−1.30	101.10	99.80	−0.0031	0.2407	−45
−40	15.261	0.01139	3.2869	−0.03	91.07	0.00	100.35	100.35	0.0000	0.2391	−40
−35	17.335	0.01145	2.9176	1.27	91.54	1.31	99.59	100.90	0.0031	0.2376	−35
−30	19.624	0.01152	2.5976	2.58	92.00	2.62	98.82	101.44	0.0061	0.2361	−30
−25	22.142	0.01159	2.3195	3.89	92.47	3.94	98.04	101.98	0.0092	0.2347	−25
−20	24.906	0.01166	2.0768	5.21	92.93	5.26	97.24	102.50	0.0122	0.2334	−20
−15	27.931	0.01173	1.8644	6.53	93.38	6.59	96.43	103.03	0.0152	0.2321	−15
−10	31.233	0.01181	1.6780	7.86	93.84	7.93	95.61	103.54	0.0182	0.2308	−10
−5	34.829	0.01188	1.5138	9.19	94.28	9.27	94.78	104.05	0.0211	0.2296	−5
0	38.734	0.01196	1.3688	10.53	94.73	10.62	93.93	104.55	0.0240	0.2284	0
5	42.967	0.01204	1.2404	11.88	95.17	11.97	93.06	105.04	0.0270	0.2272	5
10	47.545	0.01212	1.1264	13.23	95.60	13.33	92.18	105.52	0.0298	0.2261	10
15	52.486	0.01220	1.0248	14.58	96.03	14.70	91.29	105.99	0.0327	0.2250	15
20	57.808	0.01229	0.9342	15.95	96.45	16.08	90.38	106.45	0.0356	0.2240	20
25	63.529	0.01237	0.8531	17.31	96.87	17.46	89.45	106.90	0.0384	0.2230	25
30	69.668	0.01246	0.7804	18.69	97.28	18.85	88.50	107.35	0.0412	0.2220	30
35	76.245	0.01255	0.7150	20.07	97.68	20.25	87.53	107.78	0.0441	0.2210	35
40	83.278	0.01265	0.6561	21.46	98.08	21.66	86.54	108.20	0.0468	0.2200	40
45	90.787	0.01275	0.6029	22.86	98.47	23.07	85.53	108.60	0.0496	0.2191	45
50	98.792	0.01285	0.5548	24.27	98.84	24.50	84.49	108.99	0.0524	0.2182	50
55	107.31	0.01295	0.5112	25.68	99.22	25.94	83.44	109.37	0.0552	0.2173	55
60	116.37	0.01306	0.4716	27.10	99.58	27.38	82.36	109.74	0.0579	0.2164	60
65	125.98	0.01317	0.4355	28.53	99.93	28.84	81.25	110.09	0.0607	0.2155	65
70	136.18	0.01328	0.4027	29.98	100.27	30.31	80.11	110.42	0.0634	0.2147	70
75	146.97	0.01340	0.3726	31.43	100.60	31.79	78.95	110.74	0.0661	0.2138	75
80	158.38	0.01352	0.3452	32.89	100.92	33.29	77.75	111.04	0.0689	0.2130	80
85	170.44	0.01365	0.3200	34.36	101.22	34.80	76.53	111.32	0.0716	0.2121	85
90	183.16	0.01378	0.2969	35.85	101.51	36.32	75.26	111.58	0.0743	0.2113	90
95	196.57	0.01392	0.2756	37.35	101.79	37.86	73.96	111.82	0.0771	0.2104	95
100	210.69	0.01407	0.2560	38.86	102.05	39.41	72.63	112.04	0.0798	0.2095	100
105	225.54	0.01422	0.2379	40.39	102.29	40.99	71.24	112.23	0.0825	0.2087	105
110	241.15	0.01438	0.2212	41.94	102.52	42.58	69.82	112.40	0.0852	0.2078	110
115	257.55	0.01455	0.2058	43.50	102.72	44.19	68.34	112.53	0.0880	0.2069	115
120	274.75	0.01472	0.1914	45.08	102.90	45.83	66.81	112.64	0.0907	0.2060	120
140	352.17	0.01555	0.1433	51.62	103.36	52.64	60.06	112.70	0.1019	0.2021	140

R-22

Source: Tables A-7E through A-9E are calculated based on equations from A. Kamei and S. W. Beyerlein, "A Fundamental Equation for Chlorodifluoromethane (R-22)," *Fluid Phase Equilibria*, Vol. 80, No. 11, 1992, pp. 71–86.

TABLE A-8E

Properties of Saturated Refrigerant 22 (Liquid–Vapor): Pressure Table

R-22

Press. lbf/in.²	Temp. °F	Specific Volume ft³/lb		Internal Energy Btu/lb		Enthalpy Btu/lb			Entropy Btu/lb · °R		Press. lbf/in.²
		Sat. Liquid v_f	Sat. Vapor v_g	Sat. Liquid u_f	Sat. Vapor u_g	Sat. Liquid h_f	Evap. h_{fg}	Sat. Vapor h_g	Sat. Liquid s_f	Sat. Vapor s_g	
5	−78.62	0.01091	9.3014	−9.95	87.37	−9.93	105.92	95.98	−0.0248	0.2532	5
10	−55.66	0.01119	4.8769	−4.08	89.58	−4.06	102.67	98.61	−0.0098	0.2443	10
15	−40.67	0.01138	3.3402	−0.21	91.00	−0.17	100.45	100.28	−0.0004	0.2393	15
20	−29.22	0.01153	2.5518	2.78	92.07	2.83	98.70	101.52	0.0066	0.2359	20
25	−19.84	0.01166	2.0695	5.25	92.94	5.31	97.22	102.52	0.0123	0.2333	25
30	−11.82	0.01178	1.7430	7.38	93.67	7.44	95.91	103.35	0.0171	0.2313	30
35	−4.77	0.01189	1.5068	9.25	94.30	9.33	94.74	104.07	0.0212	0.2295	35
40	1.54	0.01198	1.3277	10.94	94.86	11.03	93.66	104.70	0.0249	0.2280	40
45	7.27	0.01207	1.1870	12.49	95.37	12.59	92.67	105.26	0.0283	0.2267	45
50	12.53	0.01216	1.0735	13.91	95.82	14.03	91.73	105.76	0.0313	0.2256	50
55	17.41	0.01224	0.9799	15.24	96.23	15.36	90.85	106.21	0.0341	0.2245	55
60	21.96	0.01232	0.9014	16.48	96.62	16.62	90.01	106.63	0.0367	0.2236	60
65	26.23	0.01239	0.8345	17.65	96.97	17.80	89.21	107.01	0.0391	0.2227	65
70	30.26	0.01247	0.7768	18.76	97.30	18.92	88.45	107.37	0.0414	0.2219	70
75	34.08	0.01254	0.7265	19.82	97.61	19.99	87.71	107.70	0.0435	0.2212	75
80	37.71	0.01260	0.6823	20.83	97.90	21.01	86.99	108.00	0.0456	0.2205	80
85	41.18	0.01267	0.6431	21.79	98.17	21.99	86.30	108.29	0.0475	0.2198	85
90	44.49	0.01274	0.6081	22.72	98.43	22.93	85.63	108.56	0.0494	0.2192	90
95	47.67	0.01280	0.5766	23.61	98.67	23.84	84.98	108.81	0.0511	0.2186	95
100	50.73	0.01286	0.5482	24.47	98.90	24.71	84.34	109.05	0.0528	0.2181	100
110	56.52	0.01298	0.4988	26.11	99.33	26.37	83.11	109.49	0.0560	0.2170	110
120	61.92	0.01310	0.4573	27.65	99.71	27.94	81.93	109.88	0.0590	0.2161	120
130	67.00	0.01321	0.4220	29.11	100.07	29.43	80.80	110.22	0.0618	0.2152	130
140	71.80	0.01332	0.3915	30.50	100.39	30.84	79.70	110.54	0.0644	0.2144	140
150	76.36	0.01343	0.3649	31.82	100.69	32.20	78.63	110.82	0.0669	0.2136	150
160	80.69	0.01354	0.3416	33.09	100.96	33.49	77.59	111.08	0.0693	0.2128	160
170	84.82	0.01365	0.3208	34.31	101.21	34.74	76.57	111.31	0.0715	0.2121	170
180	88.78	0.01375	0.3023	35.49	101.44	35.95	75.57	111.52	0.0737	0.2115	180
190	92.58	0.01386	0.2857	36.62	101.66	37.11	74.60	111.71	0.0757	0.2108	190
200	96.24	0.01396	0.2706	37.72	101.86	38.24	73.64	111.88	0.0777	0.2102	200
225	104.82	0.01422	0.2386	40.34	102.28	40.93	71.29	112.22	0.0824	0.2087	225
250	112.73	0.01447	0.2126	42.79	102.63	43.46	69.02	112.47	0.0867	0.2073	250
275	120.07	0.01473	0.1912	45.10	102.91	45.85	66.79	112.64	0.0908	0.2060	275
300	126.94	0.01499	0.1732	47.30	103.11	48.14	64.60	112.73	0.0946	0.2047	300
325	133.39	0.01525	0.1577	49.42	103.26	50.33	62.42	112.75	0.0982	0.2034	325
350	139.49	0.01552	0.1444	51.45	103.35	52.46	60.25	112.71	0.1016	0.2022	350

TABLE A-9E

Properties of Superheated Refrigerant 22 Vapor

T °F	v ft³/lb	u Btu/lb	h Btu/lb	s Btu/lb·°R	v ft³/lb	u Btu/lb	h Btu/lb	s Btu/lb·°R
	$p = 5$ lbf/in.² ($T_{sat} = -78.62°F$)				$p = 10$ lbf/in.² ($T_{sat} = -55.66°F$)			
Sat.	9.3014	87.37	95.98	0.2532	4.8769	89.58	98.61	0.2443
−70	9.5244	88.31	97.13	0.2562				
−60	9.7823	89.43	98.48	0.2596				
−50	10.0391	90.55	99.84	0.2630	4.9522	90.23	99.40	0.2462
−40	10.2952	91.69	101.22	0.2663	5.0846	91.39	100.81	0.2496
−30	10.5506	92.84	102.61	0.2696	5.2163	92.57	102.23	0.2530
−20	10.8054	94.01	104.01	0.2728	5.3472	93.75	103.65	0.2563
−10	11.0596	95.19	105.43	0.2760	5.4775	94.95	105.09	0.2595
0	11.3133	96.39	106.87	0.2791	5.6073	96.16	106.55	0.2627
10	11.5666	97.60	108.31	0.2822	5.7366	97.39	108.01	0.2658
20	11.8195	98.83	109.77	0.2853	5.8655	98.63	109.49	0.2690
30	12.0720	100.07	111.25	0.2884	5.9941	99.88	110.98	0.2720
40	12.3242	101.33	112.74	0.2914	6.1223	101.15	112.49	0.2751
	$p = 15$ lbf/in.² ($T_{sat} = -40.67°F$)				$p = 20$ lbf/in.² ($T_{sat} = -29.22°F$)			
Sat.	3.3402	91.00	100.28	0.2393	2.5518	92.07	101.52	0.2359
−40	3.3463	91.08	100.38	0.2396				
−30	3.4370	92.28	101.83	0.2430				
−20	3.5268	93.49	103.28	0.2463	2.6158	93.21	102.90	0.2391
−10	3.6160	94.70	104.75	0.2496	2.6846	94.45	104.39	0.2424
0	3.7046	95.93	106.22	0.2529	2.7528	95.69	105.89	0.2457
10	3.7927	97.17	107.71	0.2561	2.8204	96.95	107.39	0.2490
20	3.8804	98.43	109.20	0.2592	2.8875	98.22	108.91	0.2522
30	3.9677	99.69	110.71	0.2623	2.9542	99.49	110.43	0.2553
40	4.0546	100.97	112.23	0.2654	3.0205	100.78	111.97	0.2584
50	4.1412	102.26	113.76	0.2684	3.0865	102.09	113.52	0.2615
60	4.2275	103.57	115.31	0.2714	3.1522	103.40	115.08	0.2645
70	4.3136	104.89	116.87	0.2744	3.2176	104.73	116.65	0.2675
	$p = 25$ lbf/in.² ($T_{sat} = -19.84°F$)				$p = 30$ lbf/in.² ($T_{sat} = -11.82°F$)			
Sat.	2.0695	92.94	102.52	0.2333	1.7430	93.67	103.35	0.2313
−10	2.1252	94.18	104.02	0.2367	1.7518	93.91	103.64	0.2319
0	2.1812	95.45	105.54	0.2400	1.7997	95.19	105.19	0.2353
10	2.2365	96.72	107.07	0.2433	1.8470	96.48	106.74	0.2386
20	2.2914	98.00	108.61	0.2466	1.8937	97.78	108.30	0.2419
30	2.3458	99.29	110.15	0.2498	1.9400	99.09	109.86	0.2451
40	2.3998	100.59	111.70	0.2529	1.9858	100.40	111.43	0.2483
50	2.4535	101.91	113.27	0.2560	2.0313	101.73	113.01	0.2514
60	2.5068	103.23	114.84	0.2590	2.0764	103.06	114.60	0.2545
70	2.5599	104.57	116.42	0.2621	2.1213	104.41	116.19	0.2576
80	2.6127	105.92	118.01	0.2650	2.1659	105.77	117.80	0.2606
90	2.6654	107.28	119.62	0.2680	2.2103	107.13	119.41	0.2635
100	2.7178	108.65	121.24	0.2709	2.2545	108.52	121.04	0.2665

R-22

TABLE A-9E

(Continued)

R-22

T °F	v ft³/lb	u Btu/lb	h Btu/lb	s Btu/lb·°R	v ft³/lb	u Btu/lb	h Btu/lb	s Btu/lb·°R
	$p = 40$ lbf/in.² ($T_{sat} = 1.54°F$)				$p = 50$ lbf/in.² ($T_{sat} = 12.53°F$)			
Sat.	1.3277	94.86	104.70	0.2280	1.0735	95.82	105.76	0.2256
10	1.3593	95.99	106.06	0.2310				
20	1.3960	97.33	107.67	0.2343	1.0965	96.85	107.00	0.2282
30	1.4321	98.66	109.27	0.2376	1.1268	98.22	108.65	0.2316
40	1.4678	100.01	110.88	0.2409	1.1565	99.59	110.30	0.2349
50	1.5032	101.35	112.49	0.2441	1.1858	100.97	111.95	0.2382
60	1.5381	102.71	114.10	0.2472	1.2147	102.35	113.60	0.2414
70	1.5728	104.08	115.73	0.2503	1.2433	103.74	115.25	0.2445
80	1.6071	105.45	117.36	0.2534	1.2716	105.13	116.90	0.2476
90	1.6413	106.84	118.99	0.2564	1.2996	106.53	118.57	0.2507
100	1.6752	108.23	120.64	0.2593	1.3274	107.95	120.24	0.2537
110	1.7089	109.64	122.30	0.2623	1.3549	109.37	121.91	0.2567
120	1.7424	111.06	123.97	0.2652	1.3823	110.80	123.60	0.2596
	$p = 60$ lbf/in.² ($T_{sat} = 21.96°F$)				$p = 70$ lbf/in.² ($T_{sat} = 30.26°F$)			
Sat.	0.9014	96.62	106.63	0.2236	0.7768	97.30	107.37	0.2219
30	0.9226	97.75	108.00	0.2264				
40	0.9485	99.16	109.70	0.2298	0.7994	98.71	109.07	0.2254
50	0.9739	100.57	111.39	0.2332	0.8221	100.15	110.81	0.2288
60	0.9988	101.98	113.07	0.2365	0.8443	101.59	112.53	0.2321
70	1.0234	103.39	114.76	0.2397	0.8660	103.03	114.25	0.2354
80	1.0476	104.80	116.44	0.2428	0.8874	104.46	115.97	0.2386
90	1.0716	106.22	118.13	0.2459	0.9086	105.90	117.68	0.2418
100	1.0953	107.65	119.82	0.2490	0.9294	107.35	119.40	0.2449
110	1.1188	109.09	121.52	0.2520	0.9500	108.80	121.12	0.2479
120	1.1421	110.53	123.22	0.2549	0.9704	110.26	122.84	0.2509
130	1.1653	111.99	124.93	0.2579	0.9907	111.73	124.57	0.2539
140	1.1883	113.45	126.65	0.2608	1.0107	113.21	126.31	0.2568
	$p = 80$ lbf/in.² ($T_{sat} = 37.71°F$)				$p = 90$ lbf/in.² ($T_{sat} = 44.49°F$)			
Sat.	0.6823	97.90	108.00	0.2205	0.6081	98.43	108.56	0.2192
40	0.6871	98.24	108.42	0.2213				
50	0.7079	99.72	110.20	0.2248	0.6186	99.26	109.57	0.2212
60	0.7280	101.19	111.97	0.2283	0.6373	100.77	111.39	0.2247
70	0.7478	102.65	113.73	0.2316	0.6555	102.27	113.19	0.2282
80	0.7671	104.11	115.48	0.2349	0.6733	103.76	114.98	0.2315
90	0.7861	105.58	117.22	0.2381	0.6907	105.24	116.75	0.2348
100	0.8048	107.04	118.97	0.2412	0.7078	106.73	118.52	0.2380
110	0.8233	108.51	120.71	0.2443	0.7246	108.22	120.29	0.2411
120	0.8416	109.99	122.45	0.2474	0.7412	109.71	122.06	0.2442
130	0.8596	111.47	124.20	0.2504	0.7576	111.20	123.83	0.2472
140	0.8775	112.96	125.96	0.2533	0.7739	112.71	125.60	0.2502
150	0.8953	114.46	127.72	0.2562	0.7899	114.22	127.38	0.2531

TABLE A-9E
(Continued)

T °F	v ft³/lb	u Btu/lb	h Btu/lb	s Btu/lb·°R	v ft³/lb	u Btu/lb	h Btu/lb	s Btu/lb·°R
		p = 100 lbf/in.² (T_sat = 50.73°F)				p = 120 lbf/in.² (T_sat = 61.92°F)		
Sat.	0.5482	98.90	109.05	0.2181	0.4573	99.71	109.88	0.2161
60	0.5645	100.33	110.79	0.2214				
80	0.5980	103.38	114.46	0.2284	0.4846	102.60	113.37	0.2227
100	0.6300	106.40	118.07	0.2349	0.5130	105.73	117.13	0.2295
120	0.6609	109.42	121.66	0.2412	0.5400	108.83	120.83	0.2360
140	0.6908	112.45	125.24	0.2473	0.5661	111.92	124.50	0.2422
160	0.7201	115.50	128.83	0.2532	0.5914	115.02	128.16	0.2482
180	0.7489	118.58	132.45	0.2589	0.6161	118.15	131.84	0.2541
200	0.7771	121.69	136.08	0.2645	0.6404	121.30	135.53	0.2597
220	0.8051	124.84	139.75	0.2700	0.6642	124.48	139.24	0.2653
240	0.8327	128.04	143.45	0.2754	0.6878	127.69	142.98	0.2707
260	0.8600	131.27	147.19	0.2806	0.7110	130.95	146.75	0.2760
280	0.8871	134.54	150.97	0.2858	0.7340	134.24	150.55	0.2812
300	0.9140	137.85	154.78	0.2909	0.7568	137.57	154.39	0.2863
		p = 140 lbf/in.² (T_sat = 71.80°F)				p = 160 lbf/in.² (T_sat = 80.69°F)		
Sat.	0.3915	100.39	110.54	0.2144	0.3416	100.96	111.08	0.2128
80	0.4028	101.76	112.20	0.2175				
100	0.4289	105.02	116.14	0.2246	0.3653	104.26	115.08	0.2201
120	0.4534	108.21	119.96	0.2313	0.3881	107.56	119.06	0.2271
140	0.4768	111.37	123.73	0.2377	0.4095	110.81	122.94	0.2337
160	0.4993	114.53	127.48	0.2439	0.4301	114.03	126.77	0.2400
180	0.5212	117.70	131.21	0.2498	0.4499	117.25	130.57	0.2460
200	0.5426	120.89	134.96	0.2556	0.4692	120.47	134.37	0.2518
220	0.5636	124.10	138.71	0.2612	0.4880	123.72	138.18	0.2575
240	0.5842	127.35	142.49	0.2666	0.5065	126.99	142.00	0.2631
260	0.6045	130.62	146.30	0.2720	0.5246	130.30	145.84	0.2685
280	0.6246	133.94	150.13	0.2773	0.5425	133.63	149.70	0.2738
300	0.6445	137.29	154.00	0.2824	0.5602	137.00	153.60	0.2790
320	0.6642	140.68	157.89	0.2875	0.5777	140.41	157.62	0.2841
		p = 180 lbf/in.² (T_sat = 88.78°F)				p = 200 lbf/in.² (T_sat = 96.24°F)		
Sat.	0.3023	101.44	111.52	0.2115	0.2706	101.86	111.88	0.2102
100	0.3154	103.44	113.95	0.2159	0.2748	102.56	112.73	0.2117
120	0.3369	106.88	118.11	0.2231	0.2957	106.15	117.10	0.2194
140	0.3570	110.21	122.11	0.2299	0.3148	109.59	121.25	0.2264
160	0.3761	113.50	126.04	0.2364	0.3327	112.96	125.28	0.2330
180	0.3943	116.78	129.92	0.2425	0.3497	116.29	129.25	0.2393
200	0.4120	120.05	133.78	0.2485	0.3661	119.61	133.17	0.2454
220	0.4292	123.33	137.64	0.2542	0.3820	122.94	137.08	0.2512
240	0.4459	126.64	141.50	0.2598	0.3975	126.27	140.99	0.2569
260	0.4624	129.96	145.38	0.2653	0.4126	129.63	144.91	0.2624
280	0.4786	133.32	149.28	0.2706	0.4275	133.01	148.84	0.2678
300	0.4946	136.71	153.20	0.2759	0.4422	136.42	152.79	0.2731
320	0.5104	140.13	157.15	0.2810	0.4566	139.86	156.77	0.2782
340	0.5260	143.59	161.12	0.2860	0.4709	143.33	160.77	0.2833

R-22

TABLE A-9E

(Continued)

R-22

T °F	v ft³/lb	u Btu/lb	h Btu/lb	s Btu/lb · °R	v ft³/lb	u Btu/lb	h Btu/lb	s Btu/lb · °R
	$p = 225$ lbf/in.² ($T_{sat} = 104.82°F$)				$p = 250$ lbf/in.² ($T_{sat} = 112.73°F$)			
Sat.	0.2386	102.28	112.22	0.2087	0.2126	102.63	112.47	0.2073
120	0.2539	105.17	115.75	0.2149	0.2198	104.10	114.27	0.2104
140	0.2722	108.78	120.12	0.2223	0.2378	107.90	118.91	0.2183
160	0.2891	112.26	124.30	0.2291	0.2540	111.51	123.27	0.2255
180	0.3050	115.67	128.38	0.2356	0.2690	115.02	127.48	0.2321
200	0.3202	119.06	132.40	0.2418	0.2833	118.48	131.59	0.2385
220	0.3348	122.43	136.38	0.2477	0.2969	121.91	135.66	0.2445
240	0.3490	125.81	140.35	0.2535	0.3101	125.33	139.69	0.2504
260	0.3628	129.20	144.32	0.2591	0.3229	128.76	143.71	0.2560
280	0.3764	132.61	148.29	0.2645	0.3354	132.21	147.73	0.2616
300	0.3896	136.05	152.28	0.2699	0.3476	135.67	151.76	0.2669
320	0.4027	139.51	156.29	0.2751	0.3596	139.16	155.81	0.2722
340	0.4156	143.00	160.32	0.2802	0.3715	142.67	159.87	0.2773
360	0.4284	146.33	164.38	0.2852	0.3831	146.22	163.95	0.2824
	$p = 275$ lbf/in.² ($T_{sat} = 120.07°F$)				$p = 300$ lbf/in.² ($T_{sat} = 126.94°F$)			
Sat.	0.1912	102.91	112.64	0.2060	0.1732	103.11	112.73	0.2047
140	0.2092	106.96	117.61	0.2144	0.1849	105.93	116.20	0.2105
160	0.2250	110.73	122.19	0.2219	0.2006	109.89	121.04	0.2185
180	0.2395	144.35	126.54	0.2288	0.2146	133.64	125.56	0.2257
200	0.2530	117.88	130.77	0.2353	0.2276	117.26	129.91	0.2324
220	0.2659	121.38	134.91	0.2415	0.2399	120.83	134.15	0.2387
240	0.2782	124.85	139.02	0.2475	0.2516	124.35	138.33	0.2447
260	0.2902	128.32	143.10	0.2532	0.2629	127.87	142.47	0.2506
280	0.3018	131.80	147.17	0.2588	0.2739	131.38	146.59	0.2562
300	0.3132	135.29	151.24	0.2642	0.2845	134.90	150.71	0.2617
320	0.3243	138.80	155.32	0.2695	0.2949	138.44	154.83	0.2671
340	0.3353	142.34	159.41	0.2747	0.3051	142.00	158.95	0.2723
360	0.3461	145.90	163.53	0.2798	0.3152	145.58	163.09	0.2774
	$p = 325$ lbf/in.² ($T_{sat} = 133.39°F$)				$p = 350$ lbf/in.² ($T_{sat} = 139.49°F$)			
Sat.	0.1577	103.26	112.75	0.2034	0.1444	103.35	112.71	0.2022
140	0.1637	104.78	114.63	0.2066	0.1448	103.48	112.86	0.2024
160	0.1796	109.00	119.81	0.2151	0.1605	107.90	118.30	0.2113
180	0.1934	112.89	124.53	0.2226	0.1747	112.06	123.38	0.2194
200	0.2061	116.62	129.02	0.2295	0.1874	115.95	128.10	0.2267
220	0.2179	120.26	133.37	0.2360	0.1987	119.65	132.53	0.2333
240	0.2291	123.84	137.63	0.2422	0.2095	123.31	136.89	0.2396
260	0.2398	127.40	141.83	0.2481	0.2199	126.93	141.18	0.2457
280	0.2501	130.96	146.01	0.2538	0.2297	130.52	145.41	0.2514
300	0.2602	134.51	150.17	0.2593	0.2393	134.12	149.62	0.2571
320	0.2700	138.08	154.33	0.2647	0.2486	137.71	153.82	0.2626
340	0.2796	141.66	158.49	0.2700	0.2577	141.32	158.02	0.2679
360	0.2891	145.26	162.66	0.2752	0.2666	144.95	162.23	0.2730
380	0.2983	148.89	166.85	0.2802	0.2754	148.59	166.43	0.2781

TABLE A-10E

Properties of Saturated Refrigerant 134a (Liquid–Vapor): Temperature Table

Temp. °F	Press. lbf/in.^{2}	Specific Volume ft^3/lb		Internal Energy Btu/lb		Enthalpy Btu/lb			Entropy Btu/lb · °R		Temp. °F
		Sat. Liquid v_f	Sat. Vapor v_g	Sat. Liquid u_f	Sat. Vapor u_g	Sat. Liquid h_f	Evap. h_{fg}	Sat. Vapor h_g	Sat. Liquid s_f	Sat. Vapor s_g	
−40	7.490	0.01130	5.7173	−0.02	87.90	0.00	95.82	95.82	0.0000	0.2283	−40
−30	9.920	0.01143	4.3911	2.81	89.26	2.83	94.49	97.32	0.0067	0.2266	−30
−20	12.949	0.01156	3.4173	5.69	90.62	5.71	93.10	98.81	0.0133	0.2250	−20
−15	14.718	0.01163	3.0286	7.14	91.30	7.17	92.38	99.55	0.0166	0.2243	−15
−10	16.674	0.01170	2.6918	8.61	91.98	8.65	91.64	100.29	0.0199	0.2236	−10
−5	18.831	0.01178	2.3992	10.09	92.66	10.13	90.89	101.02	0.0231	0.2230	−5
0	21.203	0.01185	2.1440	11.58	93.33	11.63	90.12	101.75	0.0264	0.2224	0
5	23.805	0.01193	1.9208	13.09	94.01	13.14	89.33	102.47	0.0296	0.2219	5
10	26.651	0.01200	1.7251	14.60	94.68	14.66	88.53	103.19	0.0329	0.2214	10
15	29.756	0.01208	1.5529	16.13	95.35	16.20	87.71	103.90	0.0361	0.2209	15
20	33.137	0.01216	1.4009	17.67	96.02	17.74	86.87	104.61	0.0393	0.2205	20
25	36.809	0.01225	1.2666	19.22	96.69	19.30	86.02	105.32	0.0426	0.2200	25
30	40.788	0.01233	1.1474	20.78	97.35	20.87	85.14	106.01	0.0458	0.2196	30
40	49.738	0.01251	0.9470	23.94	98.67	24.05	83.34	107.39	0.0522	0.2189	40
50	60.125	0.01270	0.7871	27.14	99.98	27.28	81.46	108.74	0.0585	0.2183	50
60	72.092	0.01290	0.6584	30.39	101.27	30.56	79.49	110.05	0.0648	0.2178	60
70	85.788	0.01311	0.5538	33.68	102.54	33.89	77.44	111.33	0.0711	0.2173	70
80	101.37	0.01334	0.4682	37.02	103.78	37.27	75.29	112.56	0.0774	0.2169	80
85	109.92	0.01346	0.4312	38.72	104.39	38.99	74.17	113.16	0.0805	0.2167	85
90	118.99	0.01358	0.3975	40.42	105.00	40.72	73.03	113.75	0.0836	0.2165	90
95	128.62	0.01371	0.3668	42.14	105.60	42.47	71.86	114.33	0.0867	0.2163	95
100	138.83	0.01385	0.3388	43.87	106.18	44.23	70.66	114.89	0.0898	0.2161	100
105	149.63	0.01399	0.3131	45.62	106.76	46.01	69.42	115.43	0.0930	0.2159	105
110	161.04	0.01414	0.2896	47.39	107.33	47.81	68.15	115.96	0.0961	0.2157	110
115	173.10	0.01429	0.2680	49.17	107.88	49.63	66.84	116.47	0.0992	0.2155	115
120	185.82	0.01445	0.2481	50.97	108.42	51.47	65.48	116.95	0.1023	0.2153	120
140	243.86	0.01520	0.1827	58.39	110.41	59.08	59.57	118.65	0.1150	0.2143	140
160	314.63	0.01617	0.1341	66.26	111.97	67.20	52.58	119.78	0.1280	0.2128	160
180	400.22	0.01758	0.0964	74.83	112.77	76.13	43.78	119.91	0.1417	0.2101	180
200	503.52	0.02014	0.0647	84.90	111.66	86.77	30.92	117.69	0.1575	0.2044	200
210	563.51	0.02329	0.0476	91.84	108.48	94.27	19.18	113.45	0.1684	0.1971	210

R-134a

Source: Tables A-10E through A-12E are calculated based on equations from D. P. Wilson and R. S. Basu, "Thermodynamic Properties of a New Stratospherically Safe Working Fluid—Refrigerant 134a," *ASHRAE Trans.*, Vol. 94, Pt. 2, 1988, pp. 2095–2118.

TABLE A-11E

Properties of Saturated Refrigerant 134a (Liquid–Vapor): Pressure Table

Press. lbf/in.^{2}	Temp. °F	Specific Volume ft^3/lb		Internal Energy Btu/lb		Enthalpy Btu/lb			Entropy Btu/lb · °R		Press. lbf/in.^{2}
		Sat. Liquid v_f	Sat. Vapor v_g	Sat. Liquid u_f	Sat. Vapor u_g	Sat. Liquid h_f	Evap. h_{fg}	Sat. Vapor h_g	Sat. Liquid s_f	Sat. Vapor s_g	
5	−53.48	0.01113	8.3508	−3.74	86.07	−3.73	97.53	93.79	−0.0090	0.2311	5
10	−29.71	0.01143	4.3581	2.89	89.30	2.91	94.45	97.37	0.0068	0.2265	10
15	−14.25	0.01164	2.9747	7.36	91.40	7.40	92.27	99.66	0.0171	0.2242	15
20	−2.48	0.01181	2.2661	10.84	93.00	10.89	90.50	101.39	0.0248	0.2227	20
30	15.38	0.01209	1.5408	16.24	95.40	16.31	87.65	103.96	0.0364	0.2209	30
40	29.04	0.01232	1.1692	20.48	97.23	20.57	85.31	105.88	0.0452	0.2197	40
50	40.27	0.01252	0.9422	24.02	98.71	24.14	83.29	107.43	0.0523	0.2189	50
60	49.89	0.01270	0.7887	27.10	99.96	27.24	81.48	108.72	0.0584	0.2183	60
70	58.35	0.01286	0.6778	29.85	101.05	30.01	79.82	109.83	0.0638	0.2179	70
80	65.93	0.01302	0.5938	32.33	102.02	32.53	78.28	110.81	0.0686	0.2175	80
90	72.83	0.01317	0.5278	34.62	102.89	34.84	76.84	111.68	0.0729	0.2172	90
100	79.17	0.01332	0.4747	36.75	103.68	36.99	75.47	112.46	0.0768	0.2169	100
120	90.54	0.01360	0.3941	40.61	105.06	40.91	72.91	113.82	0.0839	0.2165	120
140	100.56	0.01386	0.3358	44.07	106.25	44.43	70.52	114.95	0.0902	0.2161	140
160	109.56	0.01412	0.2916	47.23	107.28	47.65	68.26	115.91	0.0958	0.2157	160
180	117.74	0.01438	0.2569	50.16	108.18	50.64	66.10	116.74	0.1009	0.2154	180
200	125.28	0.01463	0.2288	52.90	108.98	53.44	64.01	117.44	0.1057	0.2151	200
220	132.27	0.01489	0.2056	55.48	109.68	56.09	61.96	118.05	0.1101	0.2147	220
240	138.79	0.01515	0.1861	57.93	110.30	58.61	59.96	118.56	0.1142	0.2144	240
260	144.92	0.01541	0.1695	60.28	110.84	61.02	57.97	118.99	0.1181	0.2140	260
280	150.70	0.01568	0.1550	62.53	111.31	63.34	56.00	119.35	0.1219	0.2136	280
300	156.17	0.01596	0.1424	64.71	111.72	65.59	54.03	119.62	0.1254	0.2132	300
350	168.72	0.01671	0.1166	69.88	112.45	70.97	49.03	120.00	0.1338	0.2118	350
400	179.95	0.01758	0.0965	74.81	112.77	76.11	43.80	119.91	0.1417	0.2102	400
450	190.12	0.01863	0.0800	79.63	112.60	81.18	38.08	119.26	0.1493	0.2079	450
500	199.38	0.02002	0.0657	84.54	111.76	86.39	31.44	117.83	0.1570	0.2047	500

R-134a

TABLE A-12E

Properties of Superheated Refrigerant 134a Vapor

T °F	v ft³/lb	u Btu/lb	h Btu/lb	s Btu/lb·°R	v ft³/lb	u Btu/lb	h Btu/lb	s Btu/lb·°R
	$p = 10$ lbf/in.² ($T_{sat} = -29.71$°F)				$p = 15$ lbf/in.² ($T_{sat} = -14.25$°F)			
Sat.	4.3581	89.30	97.37	0.2265	2.9747	91.40	99.66	0.2242
-20	4.4718	90.89	99.17	0.2307				
0	4.7026	94.24	102.94	0.2391	3.0893	93.84	102.42	0.2303
20	4.9297	97.67	106.79	0.2472	3.2468	97.33	106.34	0.2386
40	5.1539	101.19	110.72	0.2553	3.4012	100.89	110.33	0.2468
60	5.3758	104.80	114.74	0.2632	3.5533	104.54	114.40	0.2548
80	5.5959	108.50	118.85	0.2709	3.7034	108.28	118.56	0.2626
100	5.8145	112.29	123.05	0.2786	3.8520	112.10	122.79	0.2703
120	6.0318	116.18	127.34	0.2861	3.9993	116.01	127.11	0.2779
140	6.2482	120.16	131.72	0.2935	4.1456	120.00	131.51	0.2854
160	6.4638	124.23	136.19	0.3009	4.2911	124.09	136.00	0.2927
180	6.6786	128.38	140.74	0.3081	4.4359	128.26	140.57	0.3000
200	6.8929	132.63	145.39	0.3152	4.5801	132.52	145.23	0.3072
	$p = 20$ lbf/in.² ($T_{sat} = -2.48$°F)				$p = 30$ lbf/in.² ($T_{sat} = 15.38$°F)			
Sat.	2.2661	93.00	101.39	0.2227	1.5408	95.40	103.96	0.2209
0	2.2816	93.43	101.88	0.2238				
20	2.4046	96.98	105.88	0.2323	1.5611	96.26	104.92	0.2229
40	2.5244	100.59	109.94	0.2406	1.6465	99.98	109.12	0.2315
60	2.6416	104.28	114.06	0.2487	1.7293	103.75	113.35	0.2398
80	2.7569	108.05	118.25	0.2566	1.8098	107.59	117.63	0.2478
100	2.8705	111.90	122.52	0.2644	1.8887	111.49	121.98	0.2558
120	2.9829	115.83	126.87	0.2720	1.9662	115.47	126.39	0.2635
140	3.0942	119.85	131.30	0.2795	2.0426	119.53	130.87	0.2711
160	3.2047	123.95	135.81	0.2869	2.1181	123.66	135.42	0.2786
180	3.3144	128.13	140.40	0.2922	2.1929	127.88	140.05	0.2859
200	3.4236	132.40	145.07	0.3014	2.2671	132.17	144.76	0.2932
220	3.5323	136.76	149.83	0.3085	2.3407	136.55	149.54	0.3003
	$p = 40$ lbf/in.² ($T_{sat} = 29.04$°F)				$p = 50$ lbf/in.² ($T_{sat} = 40.27$°F)			
Sat.	1.1692	97.23	105.88	0.2197	0.9422	98.71	107.43	0.2189
40	1.2065	99.33	108.26	0.2245				
60	1.2723	103.20	112.62	0.2331	0.9974	102.62	111.85	0.2276
80	1.3357	107.11	117.00	0.2414	1.0508	106.62	116.34	0.2361
100	1.3973	111.08	121.42	0.2494	1.1022	110.65	120.85	0.2443
120	1.4575	115.11	125.90	0.2573	1.1520	114.74	125.39	0.2523
140	1.5165	119.21	130.43	0.2650	1.2007	118.88	129.99	0.2601
160	1.5746	123.38	135.03	0.2725	1.2484	123.08	134.64	0.2677
180	1.6319	127.62	139.70	0.2799	1.2953	127.36	139.34	0.2752
200	1.6887	131.94	144.44	0.2872	1.3415	131.71	144.12	0.2825
220	1.7449	136.34	149.25	0.2944	1.3873	136.12	148.96	0.2897
240	1.8006	140.81	154.14	0.3015	1.4326	140.61	153.87	0.2969
260	1.8561	145.36	159.10	0.3085	1.4775	145.18	158.85	0.3039
280	1.9112	149.98	164.13	0.3154	1.5221	149.82	163.90	0.3108

R-134a

R-134a

TABLE A-12E

(Continued)

T °F	v ft³/lb	u Btu/lb	h Btu/lb	s Btu/lb · °R	v ft³/lb	u Btu/lb	h Btu/lb	s Btu/lb · °R
	$p = 60$ lbf/in.² ($T_{sat} = 49.89°F$)				$p = 70$ lbf/in.² ($T_{sat} = 58.35°F$)			
Sat.	0.7887	99.96	108.72	0.2183	0.6778	101.05	109.83	0.2179
60	0.8135	102.03	111.06	0.2229	0.6814	101.40	110.23	0.2186
80	0.8604	106.11	115.66	0.2316	0.7239	105.58	114.96	0.2276
100	0.9051	110.21	120.26	0.2399	0.7640	109.76	119.66	0.2361
120	0.9482	114.35	124.88	0.2480	0.8023	113.96	124.36	0.2444
140	0.9900	118.54	129.53	0.2559	0.8393	118.20	129.07	0.2524
160	1.0308	122.79	134.23	0.2636	0.8752	122.49	133.82	0.2601
180	1.0707	127.10	138.98	0.2712	0.9103	126.83	138.62	0.2678
200	1.1100	131.47	143.79	0.2786	0.9446	131.23	143.46	0.2752
220	1.1488	135.91	148.66	0.2859	0.9784	135.69	148.36	0.2825
240	1.1871	140.42	153.60	0.2930	1.0118	140.22	153.33	0.2897
260	1.2251	145.00	158.60	0.3001	1.0448	144.82	158.35	0.2968
280	1.2627	149.65	163.67	0.3070	1.0774	149.48	163.44	0.3038
300	1.3001	154.38	168.81	0.3139	1.1098	154.22	168.60	0.3107

T °F	v ft³/lb	u Btu/lb	h Btu/lb	s Btu/lb · °R	v ft³/lb	u Btu/lb	h Btu/lb	s Btu/lb · °R
	$p = 80$ lbf/in.² ($T_{sat} = 65.93°F$)				$p = 90$ lbf/in.² ($T_{sat} = 72.83°F$)			
Sat.	0.5938	102.02	110.81	0.2175	0.5278	102.89	111.68	0.2172
80	0.6211	105.03	114.23	0.2239	0.5408	104.46	113.47	0.2205
100	0.6579	109.30	119.04	0.2327	0.5751	108.82	118.39	0.2295
120	0.6927	113.56	123.82	0.2411	0.6073	113.15	123.27	0.2380
140	0.7261	117.85	128.60	0.2492	0.6380	117.50	128.12	0.2463
160	0.7584	122.18	133.41	0.2570	0.6675	121.87	132.98	0.2542
180	0.7898	126.55	138.25	0.2647	0.6961	126.28	137.87	0.2620
200	0.8205	130.98	143.13	0.2722	0.7239	130.73	142.79	0.2696
220	0.8506	135.47	148.06	0.2796	0.7512	135.25	147.76	0.2770
240	0.8803	140.02	153.05	0.2868	0.7779	139.82	152.77	0.2843
260	0.9095	144.63	158.10	0.2940	0.8043	144.45	157.84	0.2914
280	0.9384	149.32	163.21	0.3010	0.8303	149.15	162.97	0.2984
300	0.9671	154.06	168.38	0.3079	0.8561	153.91	168.16	0.3054
320	0.9955	158.88	173.62	0.3147	0.8816	158.73	173.42	0.3122

T °F	v ft³/lb	u Btu/lb	h Btu/lb	s Btu/lb · °R	v ft³/lb	u Btu/lb	h Btu/lb	s Btu/lb · °R
	$p = 100$ lbf/in.² ($T_{sat} = 79.17°F$)				$p = 120$ lbf/in.² ($T_{sat} = 90.54°F$)			
Sat.	0.4747	103.68	112.46	0.2169	0.3941	105.06	113.82	0.2165
80	0.4761	103.87	112.68	0.2173				
100	0.5086	108.32	117.73	0.2265	0.4080	107.26	116.32	0.2210
120	0.5388	112.73	122.70	0.2352	0.4355	111.84	121.52	0.2301
140	0.5674	117.13	127.63	0.2436	0.4610	116.37	126.61	0.2387
160	0.5947	121.55	132.55	0.2517	0.4852	120.89	131.66	0.2470
180	0.6210	125.99	137.49	0.2595	0.5082	125.42	136.70	0.2550
200	0.6466	130.48	142.45	0.2671	0.5305	129.97	141.75	0.2628
220	0.6716	135.02	147.45	0.2746	0.5520	134.56	146.82	0.2704
240	0.6960	139.61	152.49	0.2819	0.5731	139.20	151.92	0.2778
260	0.7201	144.26	157.59	0.2891	0.5937	143.89	157.07	0.2850
280	0.7438	148.98	162.74	0.2962	0.6140	148.63	162.26	0.2921
300	0.7672	153.75	167.95	0.3031	0.6339	153.43	167.51	0.2991
320	0.7904	158.59	173.21	0.3099	0.6537	158.29	172.81	0.3060

TABLE A-12E

(Continued)

T °F	v ft³/lb	u Btu/lb	h Btu/lb	s Btu/lb·°R	v ft³/lb	u Btu/lb	h Btu/lb	s Btu/lb·°R
	$p = 140$ lbf/in.² ($T_{sat} = 100.56°F$)				$p = 160$ lbf/in.² ($T_{sat} = 109.55°F$)			
Sat.	0.3358	106.25	114.95	0.2161	0.2916	107.28	115.91	0.2157
120	0.3610	110.90	120.25	0.2254	0.3044	109.88	118.89	0.2209
140	0.3846	115.58	125.54	0.2344	0.3269	114.73	124.41	0.2303
160	0.4066	120.21	130.74	0.2429	0.3474	119.49	129.78	0.2391
180	0.4274	124.82	135.89	0.2511	0.3666	124.20	135.06	0.2475
200	0.4474	129.44	141.03	0.2590	0.3849	128.90	140.29	0.2555
220	0.4666	134.09	146.18	0.2667	0.4023	133.61	145.52	0.2633
240	0.4852	138.77	151.34	0.2742	0.4192	138.34	150.75	0.2709
260	0.5034	143.50	156.54	0.2815	0.4356	143.11	156.00	0.2783
280	0.5212	148.28	161.78	0.2887	0.4516	147.92	161.29	0.2856
300	0.5387	153.11	167.06	0.2957	0.4672	152.78	166.61	0.2927
320	0.5559	157.99	172.39	0.3026	0.4826	157.69	171.98	0.2996
340	0.5730	162.93	177.78	0.3094	0.4978	162.65	177.39	0.3065
360	0.5898	167.93	183.21	0.3162	0.5128	167.67	182.85	0.3132
	$p = 180$ lbf/in.² ($T_{sat} = 117.74°F$)				$p = 200$ lbf/in.² ($T_{sat} = 125.28°F$)			
Sat.	0.2569	108.18	116.74	0.2154	0.2288	108.98	117.44	0.2151
120	0.2595	108.77	117.41	0.2166				
140	0.2814	113.83	123.21	0.2264	0.2446	112.87	121.92	0.2226
160	0.3011	118.74	128.77	0.2355	0.2636	117.94	127.70	0.2321
180	0.3191	123.56	134.19	0.2441	0.2809	122.88	133.28	0.2410
200	0.3361	128.34	139.53	0.2524	0.2970	127.76	138.75	0.2494
220	0.3523	133.11	144.84	0.2603	0.3121	132.60	144.15	0.2575
240	0.3678	137.90	150.15	0.2680	0.3266	137.44	149.53	0.2653
260	0.3828	142.71	155.46	0.2755	0.3405	142.30	154.90	0.2728
280	0.3974	147.55	160.79	0.2828	0.3540	147.18	160.28	0.2802
300	0.4116	152.44	166.15	0.2899	0.3671	152.10	165.69	0.2874
320	0.4256	157.38	171.55	0.2969	0.3799	157.07	171.13	0.2945
340	0.4393	162.36	177.00	0.3038	0.3926	162.07	176.60	0.3014
360	0.4529	167.40	182.49	0.3106	0.4050	167.13	182.12	0.3082
	$p = 300$ lbf/in.² ($T_{sat} = 156.17°F$)				$p = 400$ lbf/in.² ($T_{sat} = 179.95°F$)			
Sat.	0.1424	111.72	119.62	0.2132	0.0965	112.77	119.91	0.2102
160	0.1462	112.95	121.07	0.2155				
180	0.1633	118.93	128.00	0.2265	0.0965	112.79	119.93	0.2102
200	0.1777	124.47	134.34	0.2363	0.1143	120.14	128.60	0.2235
220	0.1905	129.79	140.36	0.2453	0.1275	126.35	135.79	0.2343
240	0.2021	134.99	146.21	0.2537	0.1386	132.12	142.38	0.2438
260	0.2130	140.12	151.95	0.2618	0.1484	137.65	148.64	0.2527
280	0.2234	145.23	157.63	0.2696	0.1575	143.06	154.72	0.2610
300	0.2333	150.33	163.28	0.2772	0.1660	148.39	160.67	0.2689
320	0.2428	155.44	168.92	0.2845	0.1740	153.69	166.57	0.2766
340	0.2521	160.57	174.56	0.2916	0.1816	158.97	172.42	0.2840
360	0.2611	165.74	180.23	0.2986	0.1890	164.26	178.26	0.2912
380	0.2699	170.94	185.92	0.3055	0.1962	169.57	184.09	0.2983
400	0.2786	176.18	191.64	0.3122	0.2032	174.90	189.94	0.3051

R-134a

TABLE A-13E

Properties of Saturated Ammonia (Liquid–Vapor): Temperature Table

Temp. °F	Press. lbf/in.^{2}	Specific Volume ft^3/lb		Internal Energy Btu/lb		Enthalpy Btu/lb			Entropy Btu/lb · °R		Temp. °F
		Sat. Liquid v_f	Sat. Vapor v_g	Sat. Liquid u_f	Sat. Vapor u_g	Sat. Liquid h_f	Evap. h_{fg}	Sat. Vapor h_g	Sat. Liquid s_f	Sat. Vapor s_g	
−60	5.548	0.02278	44.7537	−21.005	543.61	−20.97	610.56	589.58	−0.0512	1.4765	−60
−55	6.536	0.02288	38.3991	−15.765	545.11	−15.73	607.31	591.58	−0.0381	1.4627	−55
−50	7.664	0.02299	33.0880	−10.525	546.59	−10.49	604.04	593.54	−0.0253	1.4492	−50
−45	8.949	0.02310	28.6284	−5.295	548.04	−5.25	600.72	595.48	−0.0126	1.4361	−45
−40	10.405	0.02322	24.8672	−0.045	549.46	0.00	597.37	597.37	0.0000	1.4235	−40
−35	12.049	0.02333	21.6812	5.20	550.86	5.26	593.98	599.24	0.0124	1.4111	−35
−30	13.899	0.02345	18.9715	10.46	552.24	10.52	590.54	601.06	0.0247	1.3992	−30
−25	15.972	0.02357	16.6577	15.73	553.59	15.80	587.05	602.85	0.0369	1.3875	−25
−20	18.290	0.02369	14.6744	21.01	554.91	21.09	583.51	604.61	0.0490	1.3762	−20
−15	20.871	0.02381	12.9682	26.31	556.20	26.40	579.92	606.32	0.0610	1.3652	−15
−10	23.738	0.02393	11.4951	31.63	557.46	31.73	576.26	607.99	0.0729	1.3544	−10
−5	26.912	0.02406	10.2190	36.96	558.70	37.08	572.54	609.62	0.0847	1.3440	−5
0	30.416	0.02419	9.1100	42.32	559.91	42.45	568.76	611.22	0.0964	1.3338	0
5	34.275	0.02432	8.1430	47.69	561.08	47.85	564.92	612.76	0.1080	1.3238	5
10	38.512	0.02446	7.2974	53.09	562.23	53.27	561.00	614.27	0.1196	1.3141	10
15	43.153	0.02460	6.5556	58.52	563.34	58.72	557.01	615.73	0.1311	1.3046	15
20	48.224	0.02474	5.9032	63.97	564.43	64.19	552.95	617.14	0.1425	1.2953	20
25	53.752	0.02488	5.3278	69.43	565.48	69.68	548.82	618.51	0.1539	1.2862	25
30	59.765	0.02503	4.8188	74.93	566.49	75.20	544.62	619.82	0.1651	1.2774	30
35	66.291	0.02517	4.3675	80.44	567.48	80.75	540.34	621.09	0.1764	1.2687	35
40	73.359	0.02533	3.9664	85.98	568.42	86.33	535.97	622.30	0.1875	1.2602	40
45	81.000	0.02548	3.6090	91.55	569.33	91.93	531.54	623.46	0.1986	1.2518	45
50	89.242	0.02564	3.2897	97.13	570.21	97.55	527.02	624.57	0.2096	1.2436	50
55	98.118	0.02581	3.0040	102.73	571.04	103.20	522.42	625.62	0.2205	1.2356	55
60	107.66	0.02597	2.7476	108.35	571.83	108.87	517.74	626.61	0.2314	1.2277	60
65	117.90	0.02614	2.5171	113.99	572.59	114.56	512.97	627.54	0.2422	1.2199	65
70	128.87	0.02632	2.3095	119.65	573.29	120.28	508.12	628.40	0.2530	1.2123	70
75	140.60	0.02650	2.1220	125.33	573.95	126.02	503.18	629.20	0.2636	1.2048	75
80	153.13	0.02668	1.9524	131.02	574.57	131.78	498.15	629.93	0.2742	1.1973	80
85	166.50	0.02687	1.7988	136.73	575.13	137.56	493.03	630.59	0.2848	1.1900	85
90	180.73	0.02707	1.6593	142.46	575.65	143.37	487.81	631.18	0.2953	1.1827	90
95	195.87	0.02727	1.5324	148.21	576.10	149.20	482.49	631.68	0.3057	1.1756	95
100	211.96	0.02747	1.4168	153.98	576.51	155.05	477.06	632.11	0.3161	1.1685	100
105	229.02	0.02768	1.3113	159.76	576.85	160.94	471.52	632.46	0.3264	1.1614	105
110	247.10	0.02790	1.2149	165.58	577.13	166.85	465.86	632.71	0.3366	1.1544	110
115	266.24	0.02813	1.1266	171.41	577.34	172.80	460.08	632.88	0.3469	1.1475	115
120	286.47	0.02836	1.0456	177.28	577.48	178.79	454.16	632.95	0.3570	1.1405	120

Source: Tables A-13E through A-15E are calculated based on equations from L. Haar and J. S. Gallagher, "Thermodynamic Properties of Ammonia," *J. Phys. Chem. Reference Data*, Vol. 7, 1978, pp. 635–792.

Ammonia

TABLE A-14E

Properties of Saturated Ammonia (Liquid–Vapor): Pressure Table

Press. lbf/in.^{2}	Temp. °F	Specific Volume ft^3/lb		Internal Energy Btu/lb		Enthalpy Btu/lb			Entropy Btu/lb · °R		Press. lbf/in.^{2}
		Sat. Liquid v_f	Sat. Vapor v_g	Sat. Liquid u_f	Sat. Vapor u_g	Sat. Liquid h_f	Evap. h_{fg}	Sat. Vapor h_g	Sat. Liquid s_f	Sat. Vapor s_g	
5	−63.10	0.02271	49.320	−24.24	542.67	−24.22	612.56	588.33	−0.0593	1.4853	5
6	−57.63	0.02283	41.594	−18.51	544.32	−18.49	609.02	590.54	−0.0450	1.4699	6
7	−52.86	0.02293	36.014	−13.52	545.74	−13.49	605.92	592.42	−0.0326	1.4569	7
8	−48.63	0.02302	31.790	−9.09	546.98	−9.06	603.13	594.08	−0.0218	1.4456	8
9	−44.81	0.02311	28.477	−5.09	548.09	−5.05	600.60	595.55	−0.0121	1.4357	9
10	−41.33	0.02319	25.807	−1.44	549.09	−1.40	598.27	596.87	−0.0033	1.4268	10
12	−35.14	0.02333	21.764	5.06	550.82	5.11	594.08	599.18	0.0121	1.4115	12
14	−29.74	0.02345	18.843	10.73	552.31	10.79	590.36	601.16	0.0254	1.3986	14
16	−24.94	0.02357	16.631	15.80	553.60	15.87	587.01	602.88	0.0371	1.3874	16
18	−20.60	0.02367	14.896	20.38	554.75	20.46	583.94	604.40	0.0476	1.3775	18
20	−16.63	0.02377	13.497	24.58	555.78	24.67	581.10	605.76	0.0571	1.3687	20
25	−7.95	0.02399	10.950	33.81	557.97	33.92	574.75	608.67	0.0777	1.3501	25
30	−0.57	0.02418	9.229	41.71	559.77	41.84	569.20	611.04	0.0951	1.3349	30
35	5.89	0.02435	7.984	48.65	561.29	48.81	564.22	613.03	0.1101	1.3221	35
40	11.65	0.02450	7.041	54.89	562.60	55.07	559.69	614.76	0.1234	1.3109	40
45	16.87	0.02465	6.302	60.56	563.75	60.76	555.50	616.26	0.1354	1.3011	45
50	21.65	0.02478	5.705	65.77	564.78	66.00	551.59	617.60	0.1463	1.2923	50
55	26.07	0.02491	5.213	70.61	565.70	70.86	547.93	618.79	0.1563	1.2843	55
60	30.19	0.02503	4.801	75.13	566.53	75.41	544.46	619.87	0.1656	1.2770	60
65	34.04	0.02515	4.450	79.39	567.29	79.69	541.16	620.85	0.1742	1.2703	65
70	37.67	0.02526	4.1473	83.40	567.99	83.73	538.01	621.74	0.1823	1.2641	70
75	41.11	0.02536	3.8837	87.21	568.63	87.57	535.00	622.56	0.1900	1.2583	75
80	44.37	0.02546	3.6520	90.84	569.22	91.22	532.10	623.32	0.1972	1.2529	80
85	47.47	0.02556	3.4466	94.30	569.77	94.71	529.31	624.02	0.2040	1.2478	85
90	50.44	0.02566	3.2632	97.62	570.28	98.05	526.62	624.66	0.2106	1.2429	90
100	56.01	0.02584	2.9497	103.87	571.21	104.35	521.48	625.82	0.2227	1.2340	100
110	61.17	0.02601	2.6913	109.68	572.01	110.20	516.63	626.83	0.2340	1.2259	110
120	65.98	0.02618	2.4745	115.11	572.73	115.69	512.02	627.71	0.2443	1.2184	120
130	70.50	0.02634	2.2899	120.21	573.36	120.85	507.64	628.48	0.2540	1.2115	130
140	74.75	0.02649	2.1309	125.04	573.92	125.73	503.43	629.16	0.2631	1.2051	140
150	78.78	0.02664	1.9923	129.63	574.42	130.37	499.39	629.76	0.2717	1.1991	150
175	88.02	0.02699	1.7128	140.19	575.45	141.07	489.89	630.95	0.2911	1.1856	175
200	96.31	0.02732	1.5010	149.72	576.21	150.73	481.07	631.80	0.3084	1.1737	200
225	103.85	0.02764	1.3348	158.43	576.77	159.58	472.80	632.38	0.3240	1.1630	225
250	110.78	0.02794	1.2007	166.48	577.16	167.77	464.97	632.74	0.3382	1.1533	250
275	117.20	0.02823	1.0901	173.99	577.41	175.43	457.49	632.92	0.3513	1.1444	275
300	123.20	0.02851	0.9974	181.05	577.54	182.63	450.31	632.94	0.3635	1.1361	300

Ammonia

Ammonia *(side tab)*

TABLE A-15E

Properties of Superheated Ammonia Vapor

T °F	v ft³/lb	u Btu/lb	h Btu/lb	s Btu/lb·°R	v ft³/lb	u Btu/lb	h Btu/lb	s Btu/lb·°R
	$p = 6$ lbf/in.² ($T_{sat} = -57.63°F$)				$p = 8$ lbf/in.² ($T_{sat} = -48.63°F$)			
Sat.	41.594	544.32	590.54	1.4699	31.790	546.98	594.08	1.4456
−50	42.435	547.22	594.37	1.4793				
−40	43.533	551.03	599.40	1.4915	32.511	550.32	598.49	1.4562
−30	44.627	554.84	604.42	1.5033	33.342	554.19	603.58	1.4682
−20	45.715	558.66	609.45	1.5149	34.169	558.06	608.68	1.4799
−10	46.800	562.47	614.47	1.5261	34.992	561.93	613.76	1.4914
0	47.882	566.29	619.49	1.5372	35.811	565.79	618.84	1.5025
10	48.960	570.12	624.51	1.5480	36.627	569.66	623.91	1.5135
20	50.035	573.95	629.54	1.5586	37.440	573.52	628.99	1.5241
30	51.108	577.78	634.57	1.5690	38.250	577.40	634.06	1.5346
40	52.179	581.63	639.60	1.5791	39.058	581.27	639.13	1.5449
50	53.247	585.49	644.64	1.5891	39.865	585.16	644.21	1.5549
60	54.314	589.35	649.70	1.5990	40.669	589.05	649.29	1.5648
	$p = 10$ lbf/in.² ($T_{sat} = -41.33°F$)				$p = 12$ lbf/in.² ($T_{sat} = -35.14°F$)			
Sat.	25.807	549.09	596.87	1.4268	21.764	550.82	599.18	1.4115
−40	25.897	549.61	597.56	1.4284				
−30	26.571	553.54	602.74	1.4406	22.056	552.87	601.88	1.4178
−20	27.241	557.46	607.90	1.4525	22.621	556.85	607.12	1.4298
−10	27.906	561.37	613.05	1.4641	23.182	560.82	612.33	1.4416
0	28.568	565.29	618.19	1.4754	23.739	564.78	617.53	1.4530
10	29.227	569.19	623.31	1.4864	24.293	568.73	622.71	1.4642
20	29.882	573.10	628.43	1.4972	24.843	572.67	627.88	1.4750
30	30.535	577.01	633.55	1.5078	25.392	576.61	633.03	1.4857
40	31.186	580.91	638.66	1.5181	25.937	580.55	638.19	1.4961
50	31.835	584.82	643.77	1.5282	26.481	584.49	643.33	1.5063
60	32.482	588.74	648.89	1.5382	27.023	588.43	648.48	1.5163
70	33.127	592.66	654.01	1.5479	27.564	592.38	653.63	1.5261
	$p = 14$ lbf/in.² ($T_{sat} = -29.74°F$)				$p = 16$ lbf/in.² ($T_{sat} = -24.94°F$)			
Sat.	18.843	552.31	601.16	1.3986	16.631	553.60	602.88	1.3874
−20	19.321	556.24	606.33	1.4105	16.845	555.62	605.53	1.3935
−10	19.807	560.26	611.61	1.4223	17.275	559.69	610.88	1.4055
0	20.289	564.27	616.86	1.4339	17.701	563.75	616.19	1.4172
10	20.768	568.26	622.10	1.4452	18.124	567.79	621.48	1.4286
20	21.244	572.24	627.31	1.4562	18.544	571.81	626.75	1.4397
30	21.717	576.22	632.52	1.4669	18.961	575.82	632.00	1.4505
40	22.188	580.19	637.71	1.4774	19.376	579.82	637.23	1.4611
50	22.657	584.16	642.89	1.4877	19.789	583.82	642.45	1.4714
60	23.124	588.12	648.07	1.4977	20.200	587.81	647.66	1.4815
70	23.590	592.09	653.25	1.5076	20.609	591.80	652.86	1.4915
80	24.054	596.07	658.42	1.5173	21.017	595.80	658.07	1.5012
90	24.517	600.04	663.60	1.5268	21.424	599.80	663.27	1.5107

TABLE A-15E

(Continued)

T °F	v ft³/lb	u Btu/lb	h Btu/lb	s Btu/lb·°R	v ft³/lb	u Btu/lb	h Btu/lb	s Btu/lb·°R
	p = 18 lbf/in.² (T_{sat} = −20.60°F)				**p = 20 lbf/in.²** (T_{sat} = −16.63°F)			
Sat.	14.896	554.75	604.40	1.3775	13.497	555.78	605.76	1.3687
−20	14.919	555.00	604.72	1.3783				
−10	15.306	559.13	610.14	1.3905	13.730	558.55	609.40	1.3769
0	15.688	563.23	615.52	1.4023	14.078	562.70	614.84	1.3888
10	16.068	567.31	620.87	1.4138	14.422	566.83	620.24	1.4005
20	16.444	571.37	626.18	1.4250	14.764	570.94	625.61	1.4118
30	16.818	575.42	631.47	1.4359	15.103	575.02	630.95	1.4228
40	17.189	579.46	636.75	1.4466	15.439	579.09	636.26	1.4335
50	17.558	583.48	642.00	1.4570	15.773	583.14	641.55	1.4440
60	17.925	587.50	647.25	1.4672	16.105	587.19	646.83	1.4543
70	18.291	591.52	652.48	1.4772	16.436	591.23	652.10	1.4643
80	18.655	595.53	657.71	1.4869	16.765	595.26	657.35	1.4741
90	19.018	599.55	662.94	1.4965	17.094	599.30	662.60	1.4838
	p = 30 lbf/in.² (T_{sat} = −0.57°F)				**p = 40 lbf/in.²** (T_{sat} = 11.65°F)			
Sat.	9.2286	559.77	611.04	1.3349	7.0414	562.60	614.76	1.3109
0	9.2425	560.02	611.36	1.3356				
10	9.4834	564.38	617.07	1.3479				
20	9.7209	568.70	622.70	1.3598	7.1965	566.39	619.69	1.3213
30	9.9554	572.97	628.28	1.3713	7.3795	570.86	625.52	1.3333
40	10.187	577.21	633.80	1.3824	7.5597	575.28	631.28	1.3450
50	10.417	581.42	639.28	1.3933	7.7376	579.65	636.96	1.3562
60	10.645	585.60	644.73	1.4039	7.9134	583.97	642.58	1.3672
70	10.871	589.76	650.15	1.4142	8.0874	588.26	648.16	1.3778
80	11.096	593.90	655.54	1.4243	8.2598	592.52	653.69	1.3881
90	11.319	598.04	660.91	1.4342	8.4308	596.75	659.20	1.3982
100	11.541	602.16	666.27	1.4438	8.6006	600.97	664.67	1.4081
110	11.762	606.28	671.62	1.4533	8.7694	605.17	670.12	1.4178
	p = 50 lbf/in.² (T_{sat} = 21.65°F)				**p = 60 lbf/in.²** (T_{sat} = 30.19°F)			
Sat.	5.7049	564.78	617.60	1.2923	4.8009	566.53	619.87	1.2770
40	5.9815	573.30	628.68	1.3149	4.9278	571.25	626.00	1.2894
60	6.2733	582.31	640.39	1.3379	5.1788	580.60	638.14	1.3133
80	6.5574	591.10	651.82	1.3595	5.4218	589.66	649.90	1.3355
100	6.8358	599.75	663.04	1.3799	5.6587	598.52	661.39	1.3564
120	7.1097	608.30	674.13	1.3993	5.8910	607.23	672.68	1.3762
140	7.3802	616.80	685.13	1.4180	6.1198	615.86	683.85	1.3951
160	7.6480	625.28	696.09	1.4360	6.3458	624.44	694.95	1.4133
200	8.1776	642.27	717.99	1.4702	6.7916	641.59	717.05	1.4479
240	8.7016	659.44	740.00	1.5026	7.2318	658.87	739.21	1.4805
280	9.2218	676.88	762.26	1.5336	7.6679	676.38	761.58	1.5116
320	9.7391	694.65	784.82	1.5633	8.1013	694.21	784.22	1.5414
360	10.254	712.79	807.73	1.5919	8.5325	712.40	807.20	1.5702

Ammonia

TABLE A-15E

(Continued)

T °F	v ft³/lb	u Btu/lb	h Btu/lb	s Btu/lb · °R	v ft³/lb	u Btu/lb	h Btu/lb	s Btu/lb · °R
	p = 70 lbf/in.² (T_sat = 37.67°F)				p = 80 lbf/in.² (T_sat = 44.37°F)			
Sat.	4.1473	567.99	621.74	1.2641	3.6520	569.22	623.32	1.2529
40	4.1739	569.15	623.25	1.2671				
60	4.3962	578.85	635.84	1.2918	3.8084	577.06	633.48	1.2727
80	4.6100	588.19	647.95	1.3147	4.0006	586.69	645.95	1.2963
100	4.8175	597.26	659.70	1.3361	4.1862	595.98	657.99	1.3182
120	5.0202	606.14	671.22	1.3563	4.3668	605.04	669.73	1.3388
140	5.2193	614.91	682.56	1.3756	4.5436	613.94	681.25	1.3583
160	5.4154	623.60	693.79	1.3940	4.7175	622.74	692.63	1.3770
200	5.8015	640.91	716.11	1.4289	5.0589	640.22	715.16	1.4122
240	6.1818	658.29	738.42	1.4617	5.3942	657.71	737.62	1.4453
280	6.5580	675.89	760.89	1.4929	5.7256	675.39	760.20	1.4767
320	6.9314	693.78	783.62	1.5229	6.0540	693.34	783.02	1.5067
360	7.3026	712.02	806.67	1.5517	6.3802	711.63	806.15	1.5357
400	7.6721	730.63	830.08	1.5796	6.7047	730.29	829.61	1.5636
	p = 90 lbf/in.² (T_sat = 50.44°F)				p = 100 lbf/in.² (T_sat = 56.01°F)			
Sat.	3.2632	570.28	624.66	1.2429	2.9497	571.21	625.82	1.2340
60	3.3504	575.22	631.05	1.2553	2.9832	573.32	628.56	1.2393
80	3.5261	585.15	643.91	1.2796	3.1460	583.58	641.83	1.2644
100	3.6948	594.68	656.26	1.3021	3.3014	593.35	654.49	1.2874
120	3.8584	603.92	668.22	1.3231	3.4513	602.79	666.70	1.3088
140	4.0180	612.97	679.93	1.3430	3.5972	611.98	678.59	1.3290
160	4.1746	621.88	691.45	1.3619	3.7401	621.01	690.27	1.3481
200	4.4812	639.52	714.20	1.3974	4.0189	638.82	713.24	1.3841
240	4.7817	657.13	736.82	1.4307	4.2916	656.54	736.01	1.4176
280	5.0781	674.89	759.52	1.4623	4.5600	674.39	758.82	1.4493
320	5.3715	692.90	782.42	1.4924	4.8255	692.47	781.82	1.4796
360	5.6628	711.24	805.62	1.5214	5.0888	710.86	805.09	1.5087
400	5.9522	729.95	829.14	1.5495	5.3503	729.60	828.68	1.5368
	p = 110 lbf/in.² (T_sat = 61.17°F)				p = 120 lbf/in.² (T_sat = 65.98°F)			
Sat.	2.6913	572.01	626.83	1.2259	2.4745	572.73	627.71	1.2184
80	2.8344	581.97	639.71	1.2502	2.5744	580.33	637.53	1.2369
100	2.9791	592.00	652.69	1.2738	2.7102	590.63	650.85	1.2611
120	3.1181	601.63	665.14	1.2957	2.8401	600.46	663.57	1.2834
140	3.2528	610.98	677.24	1.3162	2.9657	609.97	675.86	1.3043
160	3.3844	620.13	689.07	1.3356	3.0879	619.24	687.86	1.3240
200	3.6406	638.11	712.27	1.3719	3.3254	637.40	711.29	1.3606
240	3.8905	655.96	735.20	1.4056	3.5563	655.36	734.39	1.3946
280	4.1362	673.88	758.13	1.4375	3.7829	673.37	757.43	1.4266
320	4.3788	692.02	781.22	1.4679	4.0065	691.58	780.61	1.4572
360	4.6192	710.47	804.56	1.4971	4.2278	710.08	804.02	1.4864
400	4.8578	729.26	828.21	1.5252	4.4473	728.92	827.74	1.5147

Ammonia

TABLE A-15E

(Continued)

T °F	v ft³/lb	u Btu/lb	h Btu/lb	s Btu/lb·°R	v ft³/lb	u Btu/lb	h Btu/lb	s Btu/lb·°R
	$p = 130$ lbf/in.² ($T_{sat} = 70.50°F$)				$p = 140$ lbf/in.² ($T_{sat} = 74.75°F$)			
Sat.	2.2899	573.36	628.48	1.2115	2.1309	573.92	629.16	1.2051
80	2.3539	578.64	635.30	1.2243	2.1633	576.80	632.89	1.2119
100	2.4824	589.23	648.98	1.2492	2.2868	587.79	647.08	1.2379
120	2.6048	599.27	661.97	1.2720	2.4004	597.85	660.08	1.2604
140	2.7226	608.94	674.48	1.2932	2.5140	607.90	673.07	1.2828
160	2.8370	618.34	686.64	1.3132	2.6204	617.34	685.27	1.3025
180	2.9488	627.57	698.55	1.3321	2.7268	626.77	697.46	1.3222
200	3.0585	636.69	710.31	1.3502	2.8289	635.93	709.27	1.3401
240	3.2734	654.77	733.57	1.3844	3.0304	654.17	732.73	1.3747
280	3.4840	672.87	756.73	1.4166	3.2274	672.38	756.04	1.4071
320	3.6915	691.14	780.00	1.4472	3.4212	690.73	779.42	1.4379
360	3.8966	709.69	803.49	1.4766	3.6126	709.34	802.99	1.4674
400	4.1000	728.57	827.27	1.5049	3.8022	728.27	826.84	1.4958
	$p = 150$ lbf/in.² ($T_{sat} = 78.78°F$)				$p = 200$ lbf/in.² ($T_{sat} = 96.31°F$)			
Sat.	1.9923	574.42	629.76	1.1991	1.5010	576.21	631.80	1.1737
100	2.1170	586.33	645.13	1.2271	1.5190	578.52	634.77	1.1790
140	2.3332	606.84	671.65	1.2729	1.6984	601.34	664.24	1.2299
180	2.5343	625.95	696.35	1.3128	1.8599	621.77	690.65	1.2726
220	2.7268	644.43	720.17	1.3489	2.0114	641.07	715.57	1.3104
260	2.9137	662.70	743.63	1.3825	2.1569	659.90	739.78	1.3450
300	3.0968	681.02	767.04	1.4141	2.2984	678.62	763.74	1.3774
340	3.2773	699.54	790.57	1.4443	2.4371	697.44	787.70	1.4081
380	3.4558	718.35	814.34	1.4733	2.5736	716.50	811.81	1.4375
420	3.6325	737.50	838.39	1.5013	2.7085	735.86	836.17	1.4659
460	3.8079	757.01	862.78	1.5284	2.8420	755.57	860.82	1.4933
500	3.9821	776.91	887.51	1.5548	2.9742	775.65	885.80	1.5199
540	4.1553	797.19	912.60	1.5804	3.1054	796.10	911.11	1.5457
580	4.3275	817.85	938.05	1.6053	3.2357	816.94	936.77	1.5709
	$p = 250$ lbf/in.² ($T_{sat} = 110.78°F$)				$p = 300$ lbf/in.² ($T_{sat} = 123.20°F$)			
Sat.	1.2007	577.16	632.74	1.1533	0.9974	577.54	632.94	1.1361
140	1.3150	595.40	656.28	1.1936	1.0568	588.94	647.65	1.1610
180	1.4539	617.38	684.69	1.2395	1.1822	612.75	678.42	1.2107
220	1.5816	637.61	710.82	1.2791	1.2944	634.01	705.91	1.2524
260	1.7025	657.03	735.85	1.3149	1.3992	654.09	731.82	1.2895
300	1.8191	676.17	760.39	1.3481	1.4994	673.69	756.98	1.3235
340	1.9328	695.32	784.79	1.3794	1.5965	693.16	781.85	1.3554
380	2.0443	714.63	809.27	1.4093	1.6913	712.74	806.70	1.3857
420	2.1540	734.22	833.93	1.4380	1.7843	732.55	831.67	1.4148
460	2.2624	754.12	858.85	1.4657	1.8759	752.66	856.87	1.4428
500	2.3695	774.38	884.07	1.4925	1.9663	773.10	882.33	1.4699
540	2.4755	795.01	909.61	1.5186	2.0556	793.90	908.09	1.4962
580	2.5807	816.01	935.47	1.5440	2.1440	815.07	934.17	1.5218

Ammonia

TABLE A-16E

Properties of Saturated Propane (Liquid–Vapor): Temperature Table

Temp. °F	Press. lbf/in.^{2}	Specific Volume ft^3/lb		Internal Energy Btu/lb		Enthalpy Btu/lb			Entropy Btu/lb · °R		Temp. °F
		Sat. Liquid v_f	Sat. Vapor v_g	Sat. Liquid u_f	Sat. Vapor u_g	Sat. Liquid h_f	Evap. h_{fg}	Sat. Vapor h_g	Sat. Liquid s_f	Sat. Vapor s_g	
−140	0.6053	0.02505	128.00	−51.33	139.22	−51.33	204.9	153.6	−0.139	0.501	−140
−120	1.394	0.02551	58.88	−41.44	143.95	−41.43	200.6	159.1	−0.109	0.481	−120
−100	2.888	0.02601	29.93	−31.34	148.80	−31.33	196.1	164.8	−0.080	0.465	−100
−80	5.485	0.02653	16.52	−21.16	153.73	−21.13	191.6	170.5	−0.053	0.452	−80
−60	9.688	0.02708	9.75	−10.73	158.74	−10.68	186.9	176.2	−0.026	0.441	−60
−40	16.1	0.02767	6.08	−0.08	163.80	0.00	181.9	181.9	0.000	0.433	−40
−20	25.4	0.02831	3.98	10.81	168.88	10.94	176.6	187.6	0.025	0.427	−20
0	38.4	0.02901	2.70	21.98	174.01	22.19	171.0	193.2	0.050	0.422	0
10	46.5	0.02939	2.25	27.69	176.61	27.94	168.0	196.0	0.063	0.420	10
20	55.8	0.02978	1.89	33.47	179.15	33.78	164.9	198.7	0.074	0.418	20
30	66.5	0.03020	1.598	39.34	181.71	39.71	161.7	201.4	0.087	0.417	30
40	78.6	0.03063	1.359	45.30	184.30	45.75	158.3	204.1	0.099	0.415	40
50	92.3	0.03110	1.161	51.36	186.74	51.89	154.7	206.6	0.111	0.414	50
60	107.7	0.03160	0.9969	57.53	189.30	58.16	151.0	209.2	0.123	0.413	60
70	124.9	0.03213	0.8593	63.81	191.71	64.55	147.0	211.6	0.135	0.412	70
80	144.0	0.03270	0.7433	70.20	194.16	71.07	142.9	214.0	0.147	0.411	80
90	165.2	0.03332	0.6447	76.72	196.46	77.74	138.4	216.2	0.159	0.410	90
100	188.6	0.03399	0.5605	83.38	198.71	84.56	133.7	218.3	0.171	0.410	100
110	214.3	0.03473	0.4881	90.19	200.91	91.56	128.7	220.3	0.183	0.409	110
120	242.5	0.03555	0.4254	97.16	202.98	98.76	123.3	222.1	0.195	0.408	120
130	273.3	0.03646	0.3707	104.33	204.92	106.17	117.5	223.7	0.207	0.406	130
140	306.9	0.03749	0.3228	111.70	206.64	113.83	111.1	225.0	0.220	0.405	140
150	343.5	0.03867	0.2804	119.33	208.05	121.79	104.1	225.9	0.233	0.403	150
160	383.3	0.04006	0.2426	127.27	209.16	130.11	96.3	226.4	0.246	0.401	160
170	426.5	0.04176	0.2085	135.60	209.81	138.90	87.4	226.3	0.259	0.398	170
180	473.4	0.04392	0.1771	144.50	209.76	148.35	76.9	225.3	0.273	0.394	180
190	524.3	0.04696	0.1470	154.38	208.51	158.94	63.8	222.8	0.289	0.387	190
200	579.7	0.05246	0.1148	166.65	204.16	172.28	44.2	216.5	0.309	0.376	200
206.1	616.1	0.07265	0.07265	186.99	186.99	195.27	0.0	195.27	0.343	0.343	206.1

TABLE A-17E

Properties of Saturated Propane (Liquid–Vapor): Pressure Table

Press. lbf/in.^{2}	Temp. °F	Specific Volume ft^3/lb		Internal Energy Btu/lb		Enthalpy Btu/lb			Entropy Btu/lb · °R		Press. lbf/in.^{2}
		Sat. Liquid v_f	Sat. Vapor v_g	Sat. Liquid u_f	Sat. Vapor u_g	Sat. Liquid h_f	Evap. h_{fg}	Sat. Vapor h_g	Sat. Liquid s_f	Sat. Vapor s_g	
0.75	−135.1	0.02516	104.8	−48.93	140.36	−48.93	203.8	154.9	−0.132	0.496	0.75
1.5	−118.1	0.02556	54.99	−40.44	144.40	−40.43	200.1	159.7	−0.106	0.479	1.5
3	−98.9	0.02603	28.9	−30.84	149.06	−30.83	196.0	165.1	−0.079	0.464	3
5	−83.0	0.02644	18.00	−22.75	152.96	−22.73	192.4	169.6	−0.057	0.454	5
7.5	−69.3	0.02682	12.36	−15.60	156.40	−15.56	189.1	173.6	−0.038	0.446	7.5
10	−58.8	0.02711	9.468	−10.10	159.04	−10.05	186.6	176.6	−0.024	0.441	10
20	−30.7	0.02796	4.971	4.93	166.18	5.03	179.5	184.6	0.012	0.430	20
30	−12.1	0.02858	3.402	15.15	170.93	15.31	174.5	189.8	0.035	0.425	30
40	2.1	0.02909	2.594	23.19	174.60	23.41	170.4	193.8	0.053	0.422	40
50	13.9	0.02954	2.099	29.96	177.63	30.23	166.8	197.1	0.067	0.419	50
60	24.1	0.02995	1.764	35.86	180.23	36.19	163.6	199.8	0.079	0.418	60
70	33.0	0.03033	1.520	41.14	182.50	41.53	160.6	202.2	0.090	0.416	70
80	41.1	0.03068	1.336	45.95	184.57	46.40	157.9	204.3	0.100	0.415	80
90	48.4	0.03102	1.190	50.38	186.36	50.90	155.3	206.2	0.109	0.414	90
100	55.1	0.03135	1.073	54.52	188.07	55.10	152.8	207.9	0.117	0.414	100
120	67.2	0.03198	0.8945	62.08	191.07	62.79	148.1	210.9	0.131	0.412	120
140	78.0	0.03258	0.7650	68.91	193.68	69.75	143.7	213.5	0.144	0.412	140
160	87.6	0.03317	0.6665	75.17	195.97	76.15	139.5	215.7	0.156	0.411	160
180	96.5	0.03375	0.5890	80.99	197.97	82.12	135.5	217.6	0.166	0.410	180
200	104.6	0.03432	0.5261	86.46	199.77	87.73	131.4	219.2	0.176	0.409	200
220	112.1	0.03489	0.4741	91.64	201.37	93.06	127.6	220.7	0.185	0.408	220
240	119.2	0.03547	0.4303	96.56	202.76	98.14	123.7	221.9	0.194	0.408	240
260	125.8	0.03606	0.3928	101.29	204.07	103.0	120.0	223.0	0.202	0.407	260
280	132.1	0.03666	0.3604	105.83	205.27	107.7	116.1	223.9	0.210	0.406	280
300	138.0	0.03727	0.3319	110.21	206.27	112.3	112.4	224.7	0.217	0.405	300
320	143.7	0.03790	0.3067	114.47	207.17	116.7	108.6	225.3	0.224	0.404	320
340	149.1	0.03855	0.2842	118.60	207.96	121.0	104.7	225.8	0.231	0.403	340
360	154.2	0.03923	0.2639	122.66	208.58	125.3	100.9	226.2	0.238	0.402	360
380	159.2	0.03994	0.2455	126.61	209.07	129.4	97.0	226.4	0.245	0.401	380
400	164.0	0.04069	0.2287	130.51	209.47	133.5	93.0	226.5	0.251	0.400	400
450	175.1	0.04278	0.1921	140.07	209.87	143.6	82.2	225.9	0.266	0.396	450
500	185.3	0.04538	0.1610	149.61	209.27	153.8	70.4	224.2	0.282	0.391	500
600	203.4	0.05659	0.1003	172.85	200.27	179.1	32.2	211.4	0.319	0.367	600
616.1	206.1	0.07265	0.07265	186.99	186.99	195.3	0.0	195.3	0.343	0.343	616.1

Propane

TABLE A-18E

Properties of Superheated Propane Vapor

T °F	v ft³/lb	u Btu/lb	h Btu/lb	s Btu/lb·°R	v ft³/lb	u Btu/lb	h Btu/lb	s Btu/lb·°R
	$p = 0.75$ lbf/in.² ($T_{sat} = -135.1$°F)				$p = 1.5$ lbf/in.² ($T_{sat} = -118.1$°F)			
Sat.	104.8	140.4	154.9	0.496	54.99	144.4	159.7	0.479
−130	106.5	141.6	156.4	0.501				
−110	113.1	146.6	162.3	0.518	56.33	146.5	162.1	0.486
−90	119.6	151.8	168.4	0.535	59.63	151.7	168.2	0.503
−70	126.1	157.2	174.7	0.551	62.92	157.1	174.5	0.520
−50	132.7	162.7	181.2	0.568	66.20	162.6	181.0	0.536
−30	139.2	168.6	187.9	0.584	69.47	168.4	187.7	0.552
−10	145.7	174.4	194.7	0.599	72.74	174.4	194.6	0.568
10	152.2	180.7	201.9	0.615	76.01	180.7	201.8	0.583
30	158.7	187.1	209.2	0.630	79.27	187.1	209.1	0.599
50	165.2	193.8	216.8	0.645	82.53	193.8	216.7	0.614
70	171.7	200.7	224.6	0.660	85.79	200.7	224.5	0.629
90	178.2	207.8	232.6	0.675	89.04	207.8	232.5	0.644
	$p = 5.0$ lbf/in.² ($T_{sat} = -83.0$°F)				$p = 10$ lbf/in.² ($T_{sat} = -58.8$°F)			
Sat.	18.00	153.0	169.6	0.454	9.468	159.0	176.6	0.441
−80	18.15	153.8	170.6	0.456				
−60	19.17	159.4	177.1	0.473				
−40	20.17	165.1	183.8	0.489	9.957	164.5	183.0	0.456
−20	21.17	171.1	190.7	0.505	10.47	170.5	190.0	0.473
0	22.17	177.2	197.7	0.521	10.98	176.7	197.1	0.489
20	23.16	183.5	205.0	0.536	11.49	183.1	204.5	0.504
40	24.15	190.1	212.5	0.552	11.99	189.7	212.0	0.520
60	25.14	196.9	220.2	0.567	12.49	196.6	219.8	0.535
80	26.13	204.0	228.2	0.582	12.99	203.6	227.8	0.550
100	27.11	211.3	236.4	0.597	13.49	210.9	236.0	0.565
120	28.09	218.8	244.8	0.611	13.99	218.5	244.4	0.580
140	29.07	226.5	253.4	0.626	14.48	226.2	253.1	0.594
	$p = 20.0$ lbf/in.² ($T_{sat} = -30.7$°F)				$p = 40.0$ lbf/in.² ($T_{sat} = 2.1$°F)			
Sat.	4.971	166.2	184.6	0.430	2.594	174.6	193.8	0.422
−20	5.117	169.5	188.5	0.439				
0	5.385	175.8	195.8	0.455				
20	5.648	182.4	203.3	0.471	2.723	180.6	200.8	0.436
40	5.909	189.1	211.0	0.487	2.864	187.6	208.8	0.453
60	6.167	195.9	218.8	0.502	3.002	194.6	216.9	0.469
80	6.424	203.1	226.9	0.518	3.137	201.8	225.1	0.484
100	6.678	210.5	235.2	0.533	3.271	209.4	233.6	0.500
120	6.932	218.0	243.7	0.548	3.403	217.0	242.2	0.515
140	7.184	225.8	252.4	0.562	3.534	224.9	251.1	0.530
160	7.435	233.9	261.4	0.577	3.664	232.9	260.1	0.545
180	7.685	242.1	270.6	0.592	3.793	241.3	269.4	0.559
200	7.935	250.6	280.0	0.606	3.921	249.8	278.9	0.574

Propane

TABLE A-18E

(*Continued*)

T °F	v ft³/lb	u Btu/lb	h Btu/lb	s Btu/lb·°R	v ft³/lb	u Btu/lb	h Btu/lb	s Btu/lb·°R
	$p = 60.0$ lbf/in.² ($T_{sat} = 24.1$°F)				$p = 80.0$ lbf/in.² ($T_{sat} = 41.1$°F)			
Sat.	1.764	180.2	199.8	0.418	1.336	184.6	204.3	0.415
30	1.794	182.4	202.3	0.384				
50	1.894	189.5	210.6	0.400	1.372	187.9	208.2	0.423
70	1.992	196.9	219.0	0.417	1.450	195.4	216.9	0.440
90	2.087	204.4	227.6	0.432	1.526	203.1	225.7	0.456
110	2.179	212.1	236.3	0.448	1.599	210.9	234.6	0.472
130	2.271	220.0	245.2	0.463	1.671	218.8	243.6	0.487
150	2.361	228.0	254.2	0.478	1.741	227.0	252.8	0.503
170	2.450	236.3	263.5	0.493	1.810	235.4	262.2	0.518
190	2.539	244.8	273.0	0.508	1.879	244.0	271.8	0.533
210	2.626	253.5	282.7	0.523	1.946	252.7	281.5	0.548
230	2.713	262.3	292.5	0.537	2.013	261.7	291.5	0.562
250	2.800	271.6	302.7	0.552	2.079	270.9	301.7	0.577
	$p = 100$ lbf/in.² ($T_{sat} = 55.1$°F)				$p = 120$ lbf/in.² ($T_{sat} = 67.2$°F)			
Sat.	1.073	188.1	207.9	0.414	0.8945	191.1	210.9	0.412
60	1.090	189.9	210.1	0.418				
80	1.156	197.8	219.2	0.435	0.9323	196.2	216.9	0.424
100	1.219	205.7	228.3	0.452	0.9887	204.3	226.3	0.441
120	1.280	213.7	237.4	0.468	1.043	212.5	235.7	0.457
140	1.340	221.9	246.7	0.483	1.094	220.8	245.1	0.473
160	1.398	230.2	256.1	0.499	1.145	229.2	254.7	0.489
180	1.454	238.8	265.7	0.514	1.194	237.9	264.4	0.504
200	1.510	247.5	275.5	0.529	1.242	246.7	274.3	0.520
220	1.566	256.4	285.4	0.544	1.289	255.6	284.3	0.534
240	1.620	265.6	295.6	0.559	1.336	264.8	294.5	0.549
260	1.674	274.9	305.9	0.573	1.382	274.2	304.9	0.564
280	1.728	284.4	316.4	0.588	1.427	283.8	315.5	0.579
	$p = 140$ lbf/in.² ($T_{sat} = 78.0$°F)				$p = 160$ lbf/in.² ($T_{sat} = 87.6$°F)			
Sat.	0.7650	193.7	213.5	0.412	0.6665	196.0	215.7	0.411
80	0.7705	213.3	214.5	0.413				
100	0.8227	222.9	224.2	0.431	0.6968	201.2	221.9	0.422
120	0.8718	232.4	233.8	0.448	0.7427	209.9	231.9	0.439
140	0.9185	242.1	243.5	0.464	0.7859	218.4	241.7	0.456
160	0.9635	251.7	253.2	0.480	0.8272	227.2	251.7	0.472
180	1.007	261.4	263.0	0.496	0.8669	235.9	261.6	0.488
200	1.050	271.4	273.0	0.511	0.9054	244.9	271.7	0.504
220	1.091	281.5	283.2	0.526	0.9430	254.0	282.0	0.519
240	1.132	291.7	293.5	0.541	0.9797	263.4	292.4	0.534
260	1.173	302.1	303.9	0.556	1.016	272.8	302.9	0.549
280	1.213	312.7	314.6	0.571	1.051	282.6	313.7	0.564
300	1.252	323.6	325.5	0.585	1.087	292.4	324.6	0.578

Propane

TABLE A-18E

(Continued)

Propane

T °F	v ft³/lb	u Btu/lb	h Btu/lb	s Btu/lb · °R	v ft³/lb	u Btu/lb	h Btu/lb	s Btu/lb · °R
	$p = 180$ lbf/in.² ($T_{sat} = 96.5°F$)				$p = 200$ lbf/in.² ($T_{sat} = 104.6°F$)			
Sat.	0.5890	198.0	217.6	0.410	0.5261	199.8	219.2	0.409
100	0.5972	199.6	219.5	0.413				
120	0.6413	208.4	229.8	0.431	0.5591	206.8	227.5	0.424
140	0.6821	217.1	239.9	0.449	0.5983	215.8	238.0	0.441
160	0.7206	226.1	250.1	0.465	0.6349	224.9	248.4	0.458
180	0.7574	234.9	260.2	0.481	0.6694	233.9	258.7	0.475
200	0.7928	244.0	270.4	0.497	0.7025	243.1	269.1	0.491
220	0.8273	253.2	280.8	0.513	0.7345	252.4	279.6	0.506
240	0.8609	262.6	291.3	0.528	0.7656	261.7	290.1	0.522
260	0.8938	272.1	301.9	0.543	0.7960	271.4	300.9	0.537
280	0.9261	281.8	312.7	0.558	0.8257	281.1	311.7	0.552
300	0.9579	291.8	323.7	0.572	0.8549	291.1	322.8	0.567
320	0.9894	301.9	334.9	0.587	0.8837	301.3	334.0	0.581
	$p = 220$ lbf/in.² ($T_{sat} = 112.1°F$)				$p = 240$ lbf/in.² ($T_{sat} = 119.2°F$)			
Sat.	0.4741	201.4	220.7	0.408	0.4303	202.8	221.9	0.408
120	0.4906	205.1	225.1	0.416	0.4321	203.2	222.4	0.409
140	0.5290	214.4	236.0	0.435	0.4704	212.9	233.8	0.428
160	0.5642	223.6	246.6	0.452	0.5048	222.4	244.8	0.446
180	0.5971	232.9	257.2	0.469	0.5365	231.6	255.5	0.463
200	0.6284	242.1	267.7	0.485	0.5664	241.1	266.3	0.480
220	0.6585	251.5	278.3	0.501	0.5949	250.5	277.0	0.496
240	0.6875	261.0	289.0	0.516	0.6223	260.1	287.8	0.511
260	0.7158	270.6	299.8	0.532	0.6490	269.8	298.7	0.527
280	0.7435	280.5	310.8	0.547	0.6749	279.8	309.8	0.542
300	0.7706	290.5	321.9	0.561	0.7002	289.8	320.9	0.557
320	0.7972	300.6	333.1	0.576	0.7251	300.1	332.3	0.571
340	0.8235	311.0	344.6	0.591	0.7496	310.5	343.8	0.586
	$p = 260$ lbf/in.² ($T_{sat} = 125.8°F$)				$p = 280$ lbf/in.² ($T_{sat} = 132.1°F$)			
Sat.	0.3928	204.1	223.0	0.407	0.3604	205.3	223.9	0.406
130	0.4012	206.3	225.6	0.411				
150	0.4374	216.1	237.2	0.431	0.3932	214.5	234.9	0.424
170	0.4697	225.8	248.4	0.449	0.4253	224.4	246.5	0.443
190	0.4995	235.2	259.3	0.466	0.4544	234.1	257.7	0.461
210	0.5275	244.8	270.2	0.482	0.4815	243.8	268.8	0.477
230	0.5541	254.4	281.1	0.498	0.5072	253.5	279.8	0.494
250	0.5798	264.2	292.1	0.514	0.5317	263.3	290.9	0.510
270	0.6046	274.1	303.2	0.530	0.5553	273.3	302.1	0.525
290	0.6288	284.0	314.3	0.545	0.5783	283.4	313.4	0.540
310	0.6524	294.3	325.7	0.560	0.6007	293.5	324.7	0.555
330	0.6756	304.7	337.2	0.574	0.6226	304.0	336.3	0.570
350	0.6984	315.2	348.8	0.589	0.6441	314.6	348.0	0.585

TABLE A-18E

(Continued)

T °F	v ft³/lb	u Btu/lb	h Btu/lb	s Btu/lb·°R	v ft³/lb	u Btu/lb	h Btu/lb	s Btu/lb·°R
				$p = 320$ lbf/in.² ($T_{sat} = 143.7°F$)				$p = 360$ lbf/in.² ($T_{sat} = 154.2°F$)
Sat.	0.3067	207.2	225.3	0.404	0.2639	208.6	226.2	0.402
150	0.3187	210.7	229.6	0.412				
170	0.3517	221.4	242.3	0.432	0.2920	217.9	237.4	0.420
190	0.3803	231.7	254.2	0.450	0.3213	228.8	250.2	0.440
210	0.4063	241.6	265.7	0.468	0.3469	239.3	262.4	0.459
230	0.4304	251.6	277.1	0.485	0.3702	249.5	274.2	0.476
250	0.4533	261.6	288.5	0.501	0.3919	259.8	285.9	0.493
270	0.4751	271.7	299.9	0.517	0.4124	270.1	297.6	0.509
290	0.4961	281.9	311.3	0.532	0.4320	280.4	309.2	0.525
310	0.5165	292.3	322.9	0.548	0.4510	290.8	320.9	0.540
330	0.5364	302.7	334.5	0.563	0.4693	301.4	332.7	0.556
350	0.5559	313.4	346.3	0.577	0.4872	312.2	344.7	0.570
370	0.5750	324.2	358.3	0.592	0.5047	323.0	356.7	0.585
				$p = 400$ lbf/in.² ($T_{sat} = 164.0°F$)				$p = 450$ lbf/in.² ($T_{sat} = 175.1°F$)
Sat.	0.2287	209.5	226.5	0.400	0.1921	209.9	225.9	0.396
170	0.2406	213.6	231.4	0.408				
190	0.2725	225.6	245.8	0.430	0.2205	220.7	239.1	0.416
210	0.2985	236.7	258.8	0.450	0.2486	233.0	253.7	0.439
230	0.3215	247.4	271.2	0.468	0.2719	244.3	267.0	0.458
250	0.3424	257.8	283.2	0.485	0.2925	255.2	279.6	0.476
270	0.3620	268.3	295.1	0.502	0.3113	266.0	292.0	0.493
290	0.3806	278.8	307.0	0.518	0.3290	276.8	304.2	0.510
310	0.3984	289.4	318.9	0.534	0.3457	287.6	316.4	0.526
330	0.4156	300.1	330.9	0.549	0.3617	298.4	328.5	0.542
350	0.4322	311.0	343.0	0.564	0.3772	309.4	340.8	0.557
370	0.4484	321.9	355.1	0.579	0.3922	320.4	353.1	0.572
390	0.4643	333.1	367.5	0.594	0.4068	331.7	365.6	0.587
				$p = 500$ lbf/in.² ($T_{sat} = 185.3°F$)				$p = 600$ lbf/in.² ($T_{sat} = 203.4°F$)
Sat.	0.1610	209.3	224.2	0.391	0.1003	200.3	211.4	0.367
190	0.1727	213.8	229.8	0.399				
210	0.2066	228.6	247.7	0.426	0.1307	214.3	228.8	0.394
230	0.2312	240.9	262.3	0.448	0.1661	232.2	250.7	0.426
250	0.2519	252.4	275.7	0.467	0.1892	245.8	266.8	0.449
270	0.2704	263.6	288.6	0.485	0.2080	258.1	281.2	0.469
290	0.2874	274.6	301.2	0.502	0.2245	269.8	294.8	0.487
310	0.3034	285.6	313.7	0.519	0.2396	281.4	308.0	0.505
330	0.3186	296.6	326.1	0.534	0.2536	292.8	321.0	0.521
350	0.3331	307.7	338.6	0.550	0.2669	304.2	333.9	0.538
370	0.3471	318.9	351.0	0.565	0.2796	315.7	346.8	0.553
390	0.3607	330.2	363.6	0.580	0.2917	327.3	359.7	0.569
410	0.3740	341.7	376.3	0.595	0.3035	338.9	372.6	0.584

Propane

TABLE A-19E

Properties of Selected Solids and Liquids: c_p, ρ, and κ

Substance	Specific Heat, c_p (Btu/lb · °R)	Density, ρ (lb/ft³)	Thermal Conductivity, κ (Btu/h · ft · °R)
Selected Solids, 540°R			
Aluminum	0.216	169	137
Coal, anthracite	0.301	84.3	0.15
Copper	0.092	557	232
Granite	0.185	164	1.61
Iron	0.107	491	46.4
Lead	0.031	705	20.4
Sand	0.191	94.9	0.16
Silver	0.056	656	248
Soil	0.439	128	0.30
Steel (AISI 302)	0.115	503	8.7
Tin	0.054	456	38.5
Building Materials, 540°R			
Brick, common	0.199	120	0.42
Concrete (stone mix)	0.210	144	0.81
Glass, plate	0.179	156	0.81
Hardboard, siding	0.279	40	0.054
Limestone	0.193	145	1.24
Plywood	0.291	34	0.069
Softwoods (fir, pine)	0.330	31.8	0.069
Insulating Materials, 540°R			
Blanket (glass fiber)	—	1.0	0.027
Cork	0.43	7.5	0.023
Duct liner (glass fiber, coated)	0.199	2.0	0.022
Polystyrene (extruded)	0.289	3.4	0.016
Vermiculite fill (flakes)	0.199	5.0	0.039
Saturated Liquids			
Ammonia, 540°R	1.151	37.5	0.269
Mercury, 540°R	0.033	845	4.94
Refrigerant 22, 540°R	0.303	74.0	0.049
Refrigerant 134a, 540°R	0.343	75.0	0.047
Unused Engine Oil, 540°R	0.456	55.2	0.084
Water, 495°R	1.006	62.42	0.332
540°R	0.998	62.23	0.354
585°R	0.999	61.61	0.373
630°R	1.002	60.79	0.386
675°R	1.008	59.76	0.394
720°R	1.017	58.55	0.398

Source: Drawn from several sources, these data are only representative. Values can vary depending on temperature, purity, moisture content, and other factors.

TABLE A-20E

Ideal Gas Specific Heats of Some Common Gases (Btu/lb · °R)

Temp. °F	c_p	c_v	k	c_p	c_v	k	c_p	c_v	k	Temp. °F
	Air			Nitrogen, N_2			Oxygen, O_2			
40	0.240	0.171	1.401	0.248	0.177	1.400	0.219	0.156	1.397	40
100	0.240	0.172	1.400	0.248	0.178	1.399	0.220	0.158	1.394	100
200	0.241	0.173	1.397	0.249	0.178	1.398	0.223	0.161	1.387	200
300	0.243	0.174	1.394	0.250	0.179	1.396	0.226	0.164	1.378	300
400	0.245	0.176	1.389	0.251	0.180	1.393	0.230	0.168	1.368	400
500	0.248	0.179	1.383	0.254	0.183	1.388	0.235	0.173	1.360	500
600	0.250	0.182	1.377	0.256	0.185	1.383	0.239	0.177	1.352	600
700	0.254	0.185	1.371	0.260	0.189	1.377	0.242	0.181	1.344	700
800	0.257	0.188	1.365	0.262	0.191	1.371	0.246	0.184	1.337	800
900	0.259	0.191	1.358	0.265	0.194	1.364	0.249	0.187	1.331	900
1000	0.263	0.195	1.353	0.269	0.198	1.359	0.252	0.190	1.326	1000
1500	0.276	0.208	1.330	0.283	0.212	1.334	0.263	0.201	1.309	1500
2000	0.286	0.217	1.312	0.293	0.222	1.319	0.270	0.208	1.298	2000

Temp. °F	c_p	c_v	k	c_p	c_v	k	c_p	c_v	k	Temp. °F
	Carbon Dioxide, CO_2			Carbon Monoxide, CO			Hydrogen, H_2			
40	0.195	0.150	1.300	0.248	0.177	1.400	3.397	2.412	1.409	40
100	0.205	0.160	1.283	0.249	0.178	1.399	3.426	2.441	1.404	100
200	0.217	0.172	1.262	0.249	0.179	1.397	3.451	2.466	1.399	200
300	0.229	0.184	1.246	0.251	0.180	1.394	3.461	2.476	1.398	300
400	0.239	0.193	1.233	0.253	0.182	1.389	3.466	2.480	1.397	400
500	0.247	0.202	1.223	0.256	0.185	1.384	3.469	2.484	1.397	500
600	0.255	0.210	1.215	0.259	0.188	1.377	3.473	2.488	1.396	600
700	0.262	0.217	1.208	0.262	0.191	1.371	3.477	2.492	1.395	700
800	0.269	0.224	1.202	0.266	0.195	1.364	3.494	2.509	1.393	800
900	0.275	0.230	1.197	0.269	0.198	1.357	3.502	2.519	1.392	900
1000	0.280	0.235	1.192	0.273	0.202	1.351	3.513	2.528	1.390	1000
1500	0.298	0.253	1.178	0.287	0.216	1.328	3.618	2.633	1.374	1500
2000	0.312	0.267	1.169	0.297	0.226	1.314	3.758	2.773	1.355	2000

Source: Adapted from K. Wark, Thermodynamics, 4th ed., McGraw-Hill, New York, 1983, as based on "Tables of Thermal Properties of Gases," NBS Circular 564, 1955.

Table A-20E

TABLE A-21E

Variation of $\bar{c}_p$ with Temperature for Selected Ideal Gases

$$\frac{\bar{c}_p}{\bar{R}} = \alpha + \beta T + \gamma T^2 + \delta T^3 + \varepsilon T^4$$

T is in °R, equations valid from 540 to 1800 °R

Gas	α	$\beta \times 10^3$	$\gamma \times 10^6$	$\delta \times 10^9$	$\varepsilon \times 10^{12}$
CO	3.710	−0.899	1.140	−0.348	0.0228
CO_2	2.401	4.853	−2.039	0.343	0
H_2	3.057	1.487	−1.793	0.947	−0.1726
H_2O	4.070	−0.616	1.281	−0.508	0.0769
O_2	3.626	−1.043	2.178	−1.160	0.2053
N_2	3.675	−0.671	0.717	−0.108	−0.0215
Air	3.653	−0.7428	1.017	−0.328	0.02632
NH_3	3.591	0.274	2.576	−1.437	0.2601
NO	4.046	−1.899	2.464	−1.048	0.1517
NO_2	3.459	1.147	2.064	−1.639	0.3448
SO_2	3.267	2.958	0.211	−0.906	0.2438
SO_3	2.578	8.087	−2.832	−0.136	0.1878
CH_4	3.826	−2.211	7.580	−3.898	0.6633
C_2H_2	1.410	10.587	−7.562	2.811	−0.3939
C_2H_4	1.426	6.324	2.466	−2.787	0.6429
Monatomic gases[a]	2.5	0	0	0	0

[a]For monatomic gases, such as He, Ne, and Ar, $\bar{c}_p$ is constant over a wide temperature range and is very nearly equal to $5/2\,\bar{R}$.

Source: Adapted from K. Wark, *Thermodynamics,* 4th ed., McGraw-Hill, New York, 1983, as based on NASA SP-273, U.S. Government Printing Office, Washington, DC, 1971.

TABLE A-22E

Ideal Gas Properties of Air

$T(°R)$, h and u(Btu/lb), $s°$(Btu/lb · °R)

				when $\Delta s = 0$[1]						when $\Delta s = 0$	
T	h	u	$s°$	p_r	v_r	T	h	u	$s°$	p_r	v_r
360	85.97	61.29	0.50369	0.3363	396.6	940	226.11	161.68	0.73509	9.834	35.41
380	90.75	64.70	0.51663	0.4061	346.6	960	231.06	165.26	0.74030	10.61	33.52
400	95.53	68.11	0.52890	0.4858	305.0	980	236.02	168.83	0.74540	11.43	31.76
420	100.32	71.52	0.54058	0.5760	270.1	1000	240.98	172.43	0.75042	12.30	30.12
440	105.11	74.93	0.55172	0.6776	240.6	1040	250.95	179.66	0.76019	14.18	27.17
460	109.90	78.36	0.56235	0.7913	215.33	1080	260.97	186.93	0.76964	16.28	24.58
480	114.69	81.77	0.57255	0.9182	193.65	1120	271.03	194.25	0.77880	18.60	22.30
500	119.48	85.20	0.58233	1.0590	174.90	1160	281.14	201.63	0.78767	21.18	20.29
520	124.27	88.62	0.59172	1.2147	158.58	1200	291.30	209.05	0.79628	24.01	18.51
537	128.34	91.53	0.59945	1.3593	146.34	1240	301.52	216.53	0.80466	27.13	16.93
540	129.06	92.04	0.60078	1.3860	144.32	1280	311.79	224.05	0.81280	30.55	15.52
560	133.86	95.47	0.60950	1.5742	131.78	1320	322.11	231.63	0.82075	34.31	14.25
580	138.66	98.90	0.61793	1.7800	120.70	1360	332.48	239.25	0.82848	38.41	13.12
600	143.47	102.34	0.62607	2.005	110.88	1400	342.90	246.93	0.83604	42.88	12.10
620	148.28	105.78	0.63395	2.249	102.12	1440	353.37	254.66	0.84341	47.75	11.17
640	153.09	109.21	0.64159	2.514	94.30	1480	363.89	262.44	0.85062	53.04	10.34
660	157.92	112.67	0.64902	2.801	87.27	1520	374.47	270.26	0.85767	58.78	9.578
680	162.73	116.12	0.65621	3.111	80.96	1560	385.08	278.13	0.86456	65.00	8.890
700	167.56	119.58	0.66321	3.446	75.25	1600	395.74	286.06	0.87130	71.73	8.263
720	172.39	123.04	0.67002	3.806	70.07	1650	409.13	296.03	0.87954	80.89	7.556
740	177.23	126.51	0.67665	4.193	65.38	1700	422.59	306.06	0.88758	90.95	6.924
760	182.08	129.99	0.68312	4.607	61.10	1750	436.12	316.16	0.89542	101.98	6.357
780	186.94	133.47	0.68942	5.051	57.20	1800	449.71	326.32	0.90308	114.0	5.847
800	191.81	136.97	0.69558	5.526	53.63	1850	463.37	336.55	0.91056	127.2	5.388
820	196.69	140.47	0.70160	6.033	50.35	1900	477.09	346.85	0.91788	141.5	4.974
840	201.56	143.98	0.70747	6.573	47.34	1950	490.88	357.20	0.92504	157.1	4.598
860	206.46	147.50	0.71323	7.149	44.57	2000	504.71	367.61	0.93205	174.0	4.258
880	211.35	151.02	0.71886	7.761	42.01	2050	518.61	378.08	0.93891	192.3	3.949
900	216.26	154.57	0.72438	8.411	39.64	2100	532.55	388.60	0.94564	212.1	3.667
920	221.18	158.12	0.72979	9.102	37.44	2150	546.54	399.17	0.95222	233.5	3.410

1. p_r and v_r data for use with Eqs. 6.41 and 6.42, respectively.

Table A-22E

TABLE A-22E
(Continued)

				$T(°R)$, h and u(Btu/lb), $s°$(Btu/lb · °R)							
				when $\Delta s = 0$[1]						when $\Delta s = 0$	
T	h	u	$s°$	p_r	v_r	T	h	u	$s°$	p_r	v_r
2200	560.59	409.78	0.95868	256.6	3.176	3700	998.11	744.48	1.10991	2330	0.5882
2250	574.69	420.46	0.96501	281.4	2.961	3750	1013.1	756.04	1.11393	2471	0.5621
2300	588.82	431.16	0.97123	308.1	2.765	3800	1028.1	767.60	1.11791	2618	0.5376
2350	603.00	441.91	0.97732	336.8	2.585	3850	1043.1	779.19	1.12183	2773	0.5143
2400	617.22	452.70	0.98331	367.6	2.419	3900	1058.1	790.80	1.12571	2934	0.4923
2450	631.48	463.54	0.98919	400.5	2.266	3950	1073.2	802.43	1.12955	3103	0.4715
2500	645.78	474.40	0.99497	435.7	2.125	4000	1088.3	814.06	1.13334	3280	0.4518
2550	660.12	485.31	1.00064	473.3	1.996	4050	1103.4	825.72	1.13709	3464	0.4331
2600	674.49	496.26	1.00623	513.5	1.876	4100	1118.5	837.40	1.14079	3656	0.4154
2650	688.90	507.25	1.01172	556.3	1.765	4150	1133.6	849.09	1.14446	3858	0.3985
2700	703.35	518.26	1.01712	601.9	1.662	4200	1148.7	860.81	1.14809	4067	0.3826
2750	717.83	529.31	1.02244	650.4	1.566	4300	1179.0	884.28	1.15522	4513	0.3529
2800	732.33	540.40	1.02767	702.0	1.478	4400	1209.4	907.81	1.16221	4997	0.3262
2850	746.88	551.52	1.03282	756.7	1.395	4500	1239.9	931.39	1.16905	5521	0.3019
2900	761.45	562.66	1.03788	814.8	1.318	4600	1270.4	955.04	1.17575	6089	0.2799
2950	776.05	573.84	1.04288	876.4	1.247	4700	1300.9	978.73	1.18232	6701	0.2598
3000	790.68	585.04	1.04779	941.4	1.180	4800	1331.5	1002.5	1.18876	7362	0.2415
3050	805.34	596.28	1.05264	1011	1.118	4900	1362.2	1026.3	1.19508	8073	0.2248
3100	820.03	607.53	1.05741	1083	1.060	5000	1392.9	1050.1	1.20129	8837	0.2096
3150	834.75	618.82	1.06212	1161	1.006	5100	1423.6	1074.0	1.20738	9658	0.1956
3200	849.48	630.12	1.06676	1242	0.9546	5200	1454.4	1098.0	1.21336	10539	0.1828
3250	864.24	641.46	1.07134	1328	0.9069	5300	1485.3	1122.0	1.21923	11481	0.1710
3300	879.02	652.81	1.07585	1418	0.8621						
3350	893.83	664.20	1.08031	1513	0.8202						
3400	908.66	675.60	1.08470	1613	0.7807						
3450	923.52	687.04	1.08904	1719	0.7436						
3500	938.40	698.48	1.09332	1829	0.7087						
3550	953.30	709.95	1.09755	1946	0.6759						
3600	968.21	721.44	1.10172	2068	0.6449						
3650	983.15	732.95	1.10584	2196	0.6157						

Ideal Gas Properties of Selected Gases

Enthalpy $\bar{h}$ (T) and internal energy $\bar{u}$ (T) in Btu/lbmol. Absolute entropy at 1 atm $\bar{s}°(T)$, in Btu/lbmol · °R.

T(°R)	Carbon Dioxide, CO_2 ($\bar{h}_f° = -169,300$ Btu/lbmol)			Carbon Monoxide, CO ($\bar{h}_f° = -47,540$ Btu/lbmol)			Water Vapor, H_2O ($\bar{h}_f° = -104,040$ Btu/lbmol)			Oxygen, O_2 ($\bar{h}_f° = 0$ Btu/lbmol)			Nitrogen, N_2 ($\bar{h}_f° = 0$ Btu/lbmol)			T(°R)
	$\bar{h}$	$\bar{u}$	$\bar{s}°$	$\bar{h}$	$\bar{u}$	$\bar{s}°$	$\bar{h}$	$\bar{u}$	$\bar{s}°$	$\bar{h}$	$\bar{u}$	$\bar{s}°$	$\bar{h}$	$\bar{u}$	$\bar{s}°$	
300	2108.2	1512.4	46.353	2081.9	1486.1	43.223	2367.6	1771.8	40.439	2073.5	1477.8	44.927	2082.0	1486.2	41.695	300
320	2256.6	1621.1	46.832	2220.9	1585.4	43.672	2526.8	1891.3	40.952	2212.6	1577.1	45.375	2221.0	1585.5	42.143	320
340	2407.3	1732.1	47.289	2359.9	1684.7	44.093	2686.0	2010.8	41.435	2351.7	1676.5	45.797	2360.0	1684.4	42.564	340
360	2560.5	1845.6	47.728	2498.8	1783.9	44.490	2845.1	2130.2	41.889	2490.8	1775.9	46.195	2498.9	1784.0	42.962	360
380	2716.4	1961.8	48.148	2637.9	1883.3	44.866	3004.4	2249.8	42.320	2630.0	1875.3	46.571	2638.0	1883.4	43.337	380
400	2874.7	2080.4	48.555	2776.9	1982.6	45.223	3163.8	2369.4	42.728	2769.1	1974.8	46.927	2777.0	1982.6	43.694	400
420	3035.7	2201.7	48.947	2916.0	2081.9	45.563	3323.2	2489.1	43.117	2908.3	2074.3	47.267	2916.1	2082.0	44.034	420
440	3199.4	2325.6	49.329	3055.0	2181.2	45.886	3482.7	2608.9	43.487	3047.5	2173.8	47.591	3055.1	2181.3	44.357	440
460	3365.7	2452.2	49.698	3194.0	2280.5	46.194	3642.3	2728.8	43.841	3186.9	2273.4	47.900	3194.1	2280.6	44.665	460
480	3534.7	2581.5	50.058	3333.0	2379.8	46.491	3802.0	2848.8	44.182	3326.5	2373.3	48.198	3333.1	2379.9	44.962	480
500	3706.2	2713.3	50.408	3472.1	2479.2	46.775	3962.0	2969.1	44.508	3466.2	2473.2	48.483	3472.2	2479.3	45.246	500
520	3880.3	2847.7	50.750	3611.2	2578.6	47.048	4122.0	3089.4	44.821	3606.1	2573.4	48.757	3611.3	2578.6	45.519	520
537	4027.5	2963.8	51.032	3725.1	2663.1	47.272	4258.0	3191.9	45.079	3725.1	2658.7	48.982	3729.5	2663.1	45.743	537
540	4056.8	2984.4	51.082	3750.3	2677.9	47.310	4282.4	3210.0	45.124	3746.2	2673.8	49.021	3750.3	2678.0	45.781	540
560	4235.8	3123.7	51.408	3889.5	2777.4	47.563	4442.8	3330.7	45.415	3886.6	2774.5	49.276	3889.5	2777.4	46.034	560
580	4417.2	3265.4	51.726	4028.7	2876.9	47.807	4603.7	3451.9	45.696	4027.3	2875.5	49.522	4028.7	2876.9	46.278	580
600	4600.9	3409.4	52.038	4168.0	2976.5	48.044	4764.7	3573.2	45.970	4168.3	2976.8	49.762	4167.9	2976.4	46.514	600
620	4786.6	3555.6	52.343	4307.4	3076.2	48.272	4926.1	3694.9	46.235	4309.7	3078.4	49.993	4307.1	3075.9	46.742	620
640	4974.9	3704.0	52.641	4446.9	3175.9	48.494	5087.8	3816.8	46.492	4451.4	3180.4	50.218	4446.4	3175.5	46.964	640
660	5165.2	3854.6	52.934	4586.6	3275.8	48.709	5250.0	3939.3	46.741	4593.5	3282.9	50.437	4585.8	3275.2	47.178	660
680	5357.6	4007.2	53.225	4726.2	3375.8	48.917	5412.5	4062.1	46.984	4736.2	3385.8	50.650	4725.3	3374.9	47.386	680
700	5552.0	4161.9	53.503	4866.0	3475.9	49.120	5575.4	4185.3	47.219	4879.3	3489.2	50.858	4864.9	3474.8	47.588	700
720	5748.4	4318.6	53.780	5006.1	3576.3	49.317	5738.8	4309.0	47.450	5022.9	3593.1	51.059	5004.5	3574.7	47.785	720
740	5946.8	4477.3	54.051	5146.4	3676.9	49.509	5902.6	4433.1	47.673	5167.0	3697.4	51.257	5144.3	3674.7	47.977	740
760	6147.0	4637.9	54.319	5286.8	3777.5	49.697	6066.9	4557.6	47.893	5311.4	3802.2	51.450	5284.1	3774.9	48.164	760
780	6349.1	4800.1	54.582	5427.4	3878.4	49.880	6231.7	4682.7	48.106	5456.4	3907.5	51.638	5424.2	3875.2	48.345	780
800	6552.9	4964.2	54.839	5568.2	3979.5	50.058	6396.9	4808.2	48.316	5602.0	4013.3	51.821	5564.4	3975.7	48.522	800
820	6758.3	5129.9	55.093	5709.4	4081.0	50.232	6562.6	4934.2	48.520	5748.1	4119.7	52.002	5704.7	4076.3	48.696	820
840	6965.7	5297.6	55.343	5850.7	4182.6	50.402	6728.9	5060.8	48.721	5894.8	4226.6	52.179	5845.3	4177.1	48.865	840
860	7174.7	5466.9	55.589	5992.3	4284.5	50.569	6895.6	5187.8	48.916	6041.9	4334.1	52.352	5985.9	4278.1	49.031	860
880	7385.3	5637.7	55.831	6134.2	4386.6	50.732	7062.9	5315.3	49.109	6189.6	4442.0	52.522	6126.9	4379.4	49.193	880
900	7597.6	5810.3	56.070	6276.4	4489.1	50.892	7230.9	5443.6	49.298	6337.9	4550.6	52.688	6268.1	4480.8	49.352	900
920	7811.4	5984.4	56.305	6419.0	4592.0	51.048	7399.4	5572.4	49.483	6486.7	4659.7	52.852	6409.6	4582.6	49.507	920
940	8026.8	6160.1	56.536	6561.7	4695.0	51.202	7568.4	5701.7	49.665	6636.1	4769.4	53.012	6551.2	4684.5	49.659	940
960	8243.8	6337.4	56.765	6704.9	4798.5	51.353	7738.0	5831.6	49.843	6786.0	4879.5	53.170	6693.1	4786.7	49.808	960
980	8462.2	6516.1	56.990	6848.4	4902.3	51.501	7908.2	5962.0	50.019	6936.4	4990.3	53.326	6835.4	4889.3	49.955	980
1000	8682.1	6696.2	57.212	6992.2	5006.3	51.646	8078.9	6093.0	50.191	7087.5	5101.6	53.477	6977.9	4992.0	50.099	1000
1020	8903.4	6877.8	57.432	7136.4	5110.8	51.788	8250.4	6224.8	50.360	7238.9	5213.3	53.628	7120.7	5095.1	50.241	1020
1040	9126.2	7060.9	57.647	7281.0	5215.7	51.929	8422.4	6357.1	50.528	7391.0	5325.7	53.775	7263.8	5198.5	50.380	1040
1060	9350.3	7245.3	57.861	7425.9	5320.9	52.067	8595.0	6490.0	50.693	7543.6	5438.6	53.921	7407.2	5302.2	50.516	1060

Table A-23E

TABLE A-23E

(Continued)

$\bar{h}$ and $\bar{u}$ in Btu/lbmol, $\bar{s}°$ in Btu/lbmol · °R

T(°R)	Carbon Dioxide, CO₂ ($\bar{h}_f° = -169{,}300$ Btu/lbmol)			Carbon Monoxide, CO ($\bar{h}_f° = -47{,}540$ Btu/lbmol)			Water Vapor, H₂O ($\bar{h}_f° = -104{,}040$ Btu/lbmol)			Oxygen, O₂ ($\bar{h}_f° = 0$ Btu/lbmol)			Nitrogen, N₂ ($\bar{h}_f° = 0$ Btu/lbmol)			T(°R)
	$\bar{h}$	$\bar{u}$	$\bar{s}°$	$\bar{h}$	$\bar{u}$	$\bar{s}°$	$\bar{h}$	$\bar{u}$	$\bar{s}°$	$\bar{h}$	$\bar{u}$	$\bar{s}°$	$\bar{h}$	$\bar{u}$	$\bar{s}°$	
1080	9575.8	7431.1	58.072	7571.1	5426.4	52.203	8768.2	6623.5	50.854	7696.8	5552.1	54.064	7551.0	5406.2	50.651	1080
1100	9802.6	7618.1	58.281	7716.8	5532.3	52.337	8942.0	6757.5	51.013	7850.4	5665.9	54.204	7695.0	5510.5	50.783	1100
1120	10030.6	7806.4	58.485	7862.9	5638.7	52.468	9116.4	6892.2	51.171	8004.5	5780.3	54.343	7839.3	5615.2	50.912	1120
1140	10260.1	7996.2	58.689	8009.2	5745.4	52.598	9291.4	7027.5	51.325	8159.1	5895.2	54.480	7984.0	5720.1	51.040	1140
1160	10490.6	8187.0	58.889	8156.1	5851.5	52.726	9467.1	7163.5	51.478	8314.2	6010.6	54.614	8129.0	5825.4	51.167	1160
1180	10722.3	8379.0	59.088	8303.3	5960.0	52.852	9643.4	7300.1	51.630	8469.8	6126.5	54.748	8274.4	5931.0	51.291	1180
1200	10955.3	8572.3	59.283	8450.8	6067.8	52.976	9820.4	7437.4	51.777	8625.8	6242.8	54.879	8420.0	6037.0	51.413	1200
1220	11189.4	8766.6	59.477	8598.8	6176.0	53.098	9998.0	7575.2	51.925	8782.4	6359.6	55.008	8566.1	6143.4	51.534	1220
1240	11424.6	8962.1	59.668	8747.2	6284.7	53.218	10176.1	7713.6	52.070	8939.4	6476.9	55.136	8712.6	6250.1	51.653	1240
1260	11661.0	9158.8	59.858	8896.0	6393.8	53.337	10354.9	7852.7	52.212	9096.7	6594.5	55.262	8859.3	6357.2	51.771	1260
1280	11898.4	9356.5	60.044	9045.0	6503.1	53.455	10534.4	7992.5	52.354	9254.6	6712.7	55.386	9006.4	6464.5	51.887	1280
1300	12136.9	9555.3	60.229	9194.6	6613.0	53.571	10714.5	8132.9	52.494	9412.9	6831.3	55.508	9153.9	6572.3	52.001	1300
1320	12376.4	9755.0	60.412	9344.6	6723.2	53.685	10895.3	8274.0	52.631	9571.6	6950.2	55.630	9301.8	6680.4	52.114	1320
1340	12617.0	9955.9	60.593	9494.8	6833.7	53.799	11076.6	8415.5	52.768	9730.7	7069.6	55.750	9450.0	6788.9	52.225	1340
1360	12858.5	10157.7	60.772	9645.5	6944.7	53.910	11258.7	8557.9	52.903	9890.2	7189.4	55.867	9598.6	6897.8	52.335	1360
1380	13101.0	10360.5	60.949	9796.6	7056.1	54.021	11441.4	8700.9	53.037	10050.1	7309.6	55.984	9747.5	7007.0	52.444	1380
1400	13344.7	10564.5	61.124	9948.1	7167.9	54.129	11624.8	8844.6	53.168	10210.4	7430.1	56.099	9896.9	7116.7	52.551	1400
1420	13589.1	10769.2	61.298	10100.0	7280.1	54.237	11808.8	8988.9	53.299	10371.0	7551.1	56.213	10046.6	7226.7	52.658	1420
1440	13834.5	10974.8	61.469	10252.2	7392.6	54.344	11993.4	9133.8	53.428	10532.0	7672.4	56.326	10196.6	7337.0	52.763	1440
1460	14080.8	11181.4	61.639	10404.8	7505.4	54.448	12178.8	9279.4	53.556	10693.3	7793.9	56.437	10347.0	7447.6	52.867	1460
1480	14328.0	11388.9	61.800	10557.8	7618.7	54.522	12364.8	9425.7	53.682	10855.1	7916.0	56.547	10497.8	7558.7	52.969	1480
1500	14576.0	11597.2	61.974	10711.1	7732.3	54.665	12551.4	9572.7	53.808	11017.1	8038.3	56.656	10648.0	7670.1	53.071	1500
1520	14824.9	11806.4	62.138	10864.9	7846.4	54.757	12738.8	9720.3	53.932	11179.6	8161.1	56.763	10800.4	7781.9	53.171	1520
1540	15074.7	12016.5	62.302	11019.0	7960.8	54.858	12926.8	9868.6	54.055	11342.4	8284.2	56.869	10952.2	7893.9	53.271	1540
1560	15325.3	12227.3	62.464	11173.4	8075.4	54.958	13115.6	10017.6	54.177	11505.4	8407.4	56.975	11104.3	8006.4	53.369	1560
1580	15576.7	12439.0	62.624	11328.2	8190.5	55.056	13305.0	10167.3	54.298	11668.8	8531.1	57.079	11256.9	8119.2	53.465	1580
1600	15829.0	12651.6	62.783	11483.4	8306.0	55.154	13494.4	10317.6	54.418	11832.5	8655.1	57.182	11409.7	8232.3	53.561	1600
1620	16081.9	12864.8	62.939	11638.9	8421.8	55.251	13685.7	10468.6	54.535	11996.6	8779.5	57.284	11562.8	8345.7	53.656	1620
1640	16335.7	13078.9	63.095	11794.7	8537.9	55.347	13877.0	10620.2	54.653	12160.9	8904.1	57.385	11716.4	8459.6	53.751	1640
1660	16590.2	13293.7	63.250	11950.9	8654.4	55.411	14069.2	10772.7	54.770	12325.5	9029.0	57.484	11870.2	8573.6	53.844	1660
1680	16845.5	13509.2	63.403	12107.5	8771.2	55.535	14261.9	10925.6	54.886	12490.4	9154.1	57.582	12024.3	8688.1	53.936	1680
1700	17101.4	13725.4	63.555	12264.3	8888.3	55.628	14455.4	11079.4	54.999	12655.6	9279.6	57.680	12178.9	8802.9	54.028	1700
1720	17358.1	13942.4	63.704	12421.4	9005.7	55.720	14649.5	11233.8	55.113	12821.1	9405.4	57.777	12333.7	8918.0	54.118	1720
1740	17615.5	14160.1	63.853	12579.0	9123.6	55.811	14844.3	11388.9	55.226	12986.9	9531.5	57.873	12488.8	9033.4	54.208	1740
1760	17873.5	14378.4	64.001	12736.7	9241.6	55.900	15039.8	11544.7	55.339	13153.0	9657.9	57.968	12644.3	9149.2	54.297	1760
1780	18132.2	14597.4	64.147	12894.9	9360.0	55.990	15236.1	11701.2	55.449	13319.2	9784.4	58.062	12800.2	9265.3	54.385	1780
1800	18391.5	14816.9	64.292	13053.2	9478.6	56.078	15433.0	11858.4	55.559	13485.8	9911.2	58.155	12956.3	9381.7	54.472	1800
1820	18651.5	15037.2	64.435	13212.0	9597.7	56.166	15630.6	12016.3	55.668	13652.5	10038.2	58.247	13112.7	9498.4	54.559	1820
1840	18912.2	15258.2	64.578	13371.0	9717.0	56.253	15828.7	12174.7	55.777	13819.6	10165.6	58.339	13269.5	9615.5	54.645	1840
1860	19173.4	15479.7	64.719	13530.2	9836.5	56.339	16027.6	12333.9	55.884	13986.8	10293.1	58.428	13426.5	9732.8	54.729	1860

(Continued)

$\bar{h}$ and $\bar{u}$ in Btu/lbmol, $\bar{s}^\circ$ in Btu/lbmol·°R

T(°R)	Carbon Dioxide, CO_2 ($\bar{h}_f^\circ = -169{,}300$ Btu/lbmol)			Carbon Monoxide, CO ($\bar{h}_f^\circ = -47{,}540$ Btu/lbmol)			Water Vapor, H_2O ($\bar{h}_f^\circ = -104{,}040$ Btu/lbmol)			Oxygen, O_2 ($\bar{h}_f^\circ = 0$ Btu/lbmol)			Nitrogen, N_2 ($\bar{h}_f^\circ = 0$ Btu/lbmol)			T(°R)
	$\bar{h}$	$\bar{u}$	$\bar{s}^\circ$	$\bar{h}$	$\bar{u}$	$\bar{s}^\circ$	$\bar{h}$	$\bar{u}$	$\bar{s}^\circ$	$\bar{h}$	$\bar{u}$	$\bar{s}^\circ$	$\bar{h}$	$\bar{u}$	$\bar{s}^\circ$	
1900	19,698	15,925	64.999	13,850	10,077	56.509	16,428	12,654	56.097	14,322	10,549	58.607	13,742	9,968	54.896	1900
1940	20,224	16,372	65.272	14,170	10,318	56.677	16,830	12,977	56.307	14,658	10,806	58.782	14,058	10,205	55.061	1940
1980	20,753	16,821	65.543	14,492	10,560	56.841	17,235	13,303	56.514	14,995	11,063	58.954	14,375	10,443	55.223	1980
2020	21,284	17,273	65.809	14,815	10,803	57.007	17,643	13,632	56.719	15,333	11,321	59.123	14,694	10,682	55.383	2020
2060	21,818	17,727	66.069	15,139	11,048	57.161	18,054	13,963	56.920	15,672	11,581	59.289	15,013	10,923	55.540	2060
2100	22,353	18,182	66.327	15,463	11,293	57.317	18,467	14,297	57.119	16,011	11,841	59.451	15,334	11,164	55.694	2100
2140	22,890	18,640	66.581	15,789	11,539	57.470	18,883	14,633	57.315	16,351	12,101	59.612	15,656	11,406	55.846	2140
2180	23,429	19,101	66.830	16,116	11,787	57.621	19,301	14,972	57.509	16,692	12,363	59.770	15,978	11,649	55.995	2180
2220	23,970	19,561	67.076	16,443	12,035	57.770	19,722	15,313	57.701	17,036	12,625	59.926	16,302	11,893	56.141	2220
2260	24,512	20,024	67.319	16,722	12,284	57.917	20,145	15,657	57.889	17,376	12,888	60.077	16,626	12,138	56.286	2260
2300	25,056	20,489	67.557	17,101	12,534	58.062	20,571	16,003	58.077	17,719	13,151	60.228	16,951	12,384	56.429	2300
2340	25,602	20,955	67.792	17,431	12,784	58.204	20,999	16,352	58.261	18,062	13,416	60.376	17,277	12,630	56.570	2340
2380	26,150	21,423	68.025	17,762	13,035	58.344	21,429	16,703	58.445	18,407	13,680	60.522	17,604	12,878	56.708	2380
2420	26,699	21,893	68.253	18,093	13,287	58.482	21,862	17,057	58.625	18,752	13,946	60.666	17,932	13,126	56.845	2420
2460	27,249	22,364	68.479	18,426	13,541	58.619	22,298	17,413	58.803	19,097	14,212	60.808	18,260	13,375	56.980	2460
2500	27,801	22,837	68.702	18,759	13,794	58.754	22,735	17,771	58.980	19,443	14,479	60.946	18,590	13,625	57.112	2500
2540	28,355	23,310	68.921	19,093	14,048	58.885	23,175	18,131	59.155	19,790	14,746	61.084	18,919	13,875	57.243	2540
2580	28,910	23,786	69.138	19,427	14,303	59.016	23,618	18,494	59.328	20,138	15,014	61.220	19,250	14,127	57.372	2580
2620	29,465	24,262	69.352	19,762	14,559	59.145	24,062	18,859	59.500	20,485	15,282	61.354	19,582	14,379	57.499	2620
2660	30,023	24,740	69.563	20,098	14,815	59.272	24,508	19,226	59.669	20,834	15,551	61.486	19,914	14,631	57.625	2660
2700	30,581	25,220	69.771	20,434	15,072	59.398	24,957	19,595	59.837	21,183	15,821	61.616	20,246	14,885	57.750	2700
2740	31,141	25,701	69.977	20,771	15,330	59.521	25,408	19,967	60.003	21,533	16,091	61.744	20,580	15,139	57.872	2740
2780	31,702	26,181	70.181	21,108	15,588	59.644	25,861	20,340	60.167	21,883	16,362	61.871	20,914	15,393	57.993	2780
2820	32,264	26,664	70.382	21,446	15,846	59.765	26,316	20,715	60.330	22,232	16,633	61.996	21,248	15,648	58.113	2820
2860	32,827	27,148	70.580	21,785	16,105	59.884	26,773	21,093	60.490	22,584	16,905	62.120	21,584	15,905	58.231	2860
2900	33,392	27,633	70.776	22,124	16,365	60.002	27,231	21,472	60.650	22,936	17,177	62.242	21,920	16,161	58.348	2900
2940	33,957	28,118	70.970	22,463	16,625	60.118	27,692	21,853	60.809	23,288	17,450	62.363	22,256	16,417	58.463	2940
2980	34,523	28,605	71.160	22,803	16,885	60.232	28,154	22,237	60.965	23,641	17,723	62.483	22,593	16,675	58.576	2980
3020	35,090	29,093	71.350	23,144	17,146	60.346	28,619	22,621	61.120	23,994	17,997	62.599	22,930	16,933	58.688	3020
3060	35,659	29,582	71.537	23,485	17,408	60.458	29,085	23,085	61.274	24,348	18,271	62.716	23,268	17,192	58.800	3060
3100	36,228	30,072	71.722	23,826	17,670	60.569	29,553	23,397	61.426	24,703	18,546	62.831	23,607	17,451	58.910	3100
3140	36,798	30,562	71.904	24,168	17,932	60.679	30,023	23,787	61.577	25,057	18,822	62.945	23,946	17,710	59.019	3140
3180	37,369	31,054	72.085	24,510	18,195	60.787	30,494	24,179	61.727	25,413	19,098	63.057	24,285	17,970	59.126	3180
3220	37,941	31,546	72.264	24,853	18,458	60.894	30,967	24,572	61.874	25,769	19,374	63.169	24,625	18,231	59.232	3220
3260	38,513	32,039	72.441	25,196	18,722	61.000	31,442	24,968	62.022	26,175	19,651	63.279	24,965	18,491	59.338	3260
3300	39,087	32,533	72.616	25,539	18,986	61.105	31,918	25,365	62.167	26,412	19,928	63.386	25,306	18,753	59.442	3300
3340	39,661	33,028	72.788	25,883	19,250	61.209	32,396	25,763	62.312	26,839	20,206	63.494	25,647	19,014	59.544	3340
3380	40,236	33,524	72.960	26,227	19,515	61.311	32,876	26,164	62.454	27,197	20,485	63.601	25,989	19,277	59.646	3380
3420	40,812	34,020	73.129	26,572	19,780	61.412	33,357	26,565	62.597	27,555	20,763	63.706	26,331	19,539	59.747	3420
3460	41,338	34,517	73.297	26,917	20,045	61.513	33,839	26,968	62.738	27,914	21,043	63.811	26,673	19,802	59.846	3460

Table A-23E

TABLE A-23E

(Continued)

$\bar{h}$ and $\bar{u}$ in Btu/lbmol, $\bar{s}°$ in Btu/lbmol · °R

$T(°R)$	Carbon Dioxide, CO_2 ($\bar{h}_f° = -169,300$ Btu/lbmol)			Carbon Monoxide, CO ($\bar{h}_f° = -47,540$ Btu/lbmol)			Water Vapor, H_2O ($\bar{h}_f° = -104,040$ Btu/lbmol)			Oxygen, O_2 ($\bar{h}_f° = 0$ Btu/lbmol)			Nitrogen, N_2 ($\bar{h}_f° = 0$ Btu/lbmol)			$T(°R)$
	$\bar{h}$	$\bar{u}$	$\bar{s}°$	$\bar{h}$	$\bar{u}$	$\bar{s}°$	$\bar{h}$	$\bar{u}$	$\bar{s}°$	$\bar{h}$	$\bar{u}$	$\bar{s}°$	$\bar{h}$	$\bar{u}$	$\bar{s}°$	
3500	41,965	35,015	73.462	27,262	20,311	61.612	34,324	27,373	62.875	28,273	21,323	63.914	27,016	20,065	59.944	3500
3540	42,543	35,513	73.627	27,608	20,576	61.710	34,809	27,779	63.015	28,633	21,603	64.016	27,359	20,329	60.041	3540
3580	43,121	36,012	73.789	27,954	20,844	61.807	35,296	28,187	63.153	28,994	21,884	64.114	27,703	20,593	60.138	3580
3620	43,701	36,512	73.951	28,300	21,111	61.903	35,785	28,596	63.288	29,354	22,165	64.217	28,046	20,858	60.230	3620
3660	44,280	37,012	74.110	28,647	21,378	61.998	36,274	29,006	63.423	29,716	22,447	64.316	28,391	21,122	60.328	3660
3700	44,861	37,513	74.267	28,994	21,646	62.093	36,765	29,418	63.557	30,078	22,730	64.415	28,735	21,387	60.422	3700
3740	45,442	38,014	74.423	29,341	21,914	62.186	37,258	29,831	63.690	30,440	23,013	64.512	29,080	21,653	60.515	3740
3780	46,023	38,517	74.578	29,688	22,182	62.279	37,752	30,245	63.821	30,803	23,296	64.609	29,425	21,919	60.607	3780
3820	46,605	39,019	74.732	30,036	22,450	62.370	38,247	30,661	63.952	31,166	23,580	64.704	29,771	22,185	60.698	3820
3860	47,188	39,522	74.884	30,384	22,719	62.461	38,743	31,077	64.082	31,529	23,864	64.800	30,117	22,451	60.788	3860
3900	47,771	40,026	75.033	30,733	22,988	62.551	39,240	31,495	64.210	31,894	24,149	64.893	30,463	22,718	60.877	3900
3940	48,355	40,531	75.182	31,082	23,257	62.640	39,739	31,915	64.338	32,258	24,434	64.986	30,809	22,985	60.966	3940
3980	48,939	41,035	75.330	31,431	23,527	62.728	40,239	32,335	64.465	32,623	24,720	65.078	31,156	23,252	61.053	3980
4020	49,524	41,541	75.477	31,780	23,797	62.816	40,740	32,757	64.591	32,989	25,006	65.169	31,503	23,520	61.139	4020
4060	50,109	42,047	75.622	32,129	24,067	62.902	41,242	33,179	64.715	33,355	25,292	65.260	31,850	23,788	61.225	4060
4100	50,695	42,553	75.765	32,479	24,337	62.988	41,745	33,603	64.839	33,722	25,580	65.350	32,198	24,056	61.310	4100
4140	51,282	43,060	75.907	32,829	24,608	63.072	42,250	34,028	64.962	34,089	25,867	65.439	32,546	24,324	61.395	4140
4180	51,868	43,568	76.048	33,179	24,878	63.156	42,755	34,454	65.084	34,456	26,155	65.527	32,894	24,593	61.479	4180
4220	52,456	44,075	76.188	33,530	25,149	63.240	43,267	34,881	65.204	34,824	26,444	65.615	33,242	24,862	61.562	4220
4260	53,044	44,584	76.327	33,880	25,421	63.323	43,769	35,310	65.325	35,192	26,733	65.702	33,591	25,131	61.644	4260
4300	53,632	45,093	76.464	34,231	25,692	63.405	44,278	35,739	65.444	35,561	27,022	65.788	33,940	25,401	61.726	4300
4340	54,221	45,602	76.601	34,582	25,964	63.486	44,788	36,169	65.563	35,930	27,312	65.873	34,289	25,670	61.806	4340
4380	54,810	46,112	76.736	34,934	26,235	63.567	45,298	36,600	65.680	36,300	27,602	65.958	34,638	25,940	61.887	4380
4420	55,400	46,622	76.870	35,285	26,508	63.647	45,810	37,032	65.797	36,670	27,823	66.042	34,988	26,210	61.966	4420
4460	55,990	47,133	77.003	35,637	26,780	63.726	46,322	37,465	65.913	37,041	28,184	66.125	35,338	26,481	62.045	4460
4500	56,581	47,645	77.135	35,989	27,052	63.805	46,836	37,900	66.028	37,412	28,475	66.208	35,688	26,751	62.123	4500
4540	57,172	48,156	77.266	36,341	27,325	63.883	47,350	38,334	66.142	37,783	28,768	66.290	36,038	27,022	62.201	4540
4580	57,764	48,668	77.395	36,693	27,598	63.960	47,866	38,770	66.255	38,155	29,060	66.372	36,389	27,293	62.278	4580
4620	58,356	49,181	77.581	37,046	27,871	64.036	48,382	39,207	66.368	38,528	29,353	66.453	36,739	27,565	62.354	4620
4660	58,948	49,694	77.652	37,398	28,144	64.113	48,899	39,645	66.480	38,900	29,646	66.533	37,090	27,836	62.429	4660
4700	59,541	50,208	77.779	37,751	28,417	64.188	49,417	40,083	66.591	39,274	29,940	66.613	37,441	28,108	62.504	4700
4740	60,134	50,721	77.905	38,104	28,691	64.263	49,936	40,523	66.701	39,647	30,234	66.691	37,792	28,379	62.578	4740
4780	60,728	51,236	78.029	38,457	28,965	64.337	50,455	40,963	66.811	40,021	30,529	66.770	38,144	28,651	62.652	4780
4820	61,322	51,750	78.153	38,811	29,239	64.411	50,976	41,404	66.920	40,396	30,824	66.848	38,495	28,924	62.725	4820
4860	61,916	52,265	78.276	39,164	29,513	64.484	51,497	41,856	67.028	40,771	31,120	66.925	38,847	29,196	62.798	4860
4900	62,511	52,781	78.398	39,518	29,787	64.556	52,019	42,288	67.135	41,146	31,415	67.003	39,199	29,468	62.870	4900
5000	64,000	54,071	78.698	40,403	30,473	64.735	53,327	43,398	67.401	42,086	32,157	67.193	40,080	30,151	63.049	5000
5100	65,491	55,363	78.994	41,289	31,161	64.910	54,640	44,512	67.662	43,021	32,901	67.380	40,962	30,834	63.223	5100
5200	66,984	56,658	79.284	42,176	31,849	65.082	55,957	45,631	67.918	43,974	33,648	67.562	41,844	31,518	63.395	5200
5300	68,471	57,954	79.569	43,063	32,538	65.252	57,279	46,754	68.172	44,922	34,397	67.743	42,728	32,203	63.563	5300

TABLE A-24E

Constants for the van der Waals, Redlich–Kwong, and Benedict–Webb–Rubin Equations of State

1. van der Waals and Redlich–Kwong: Constants for pressure in atm, specific volume in $ft^3/lbmol$, and temperature in °R

| Substance | van der Waals | | Redlich–Kwong | |
	$atm\left(\dfrac{ft^3}{lbmol}\right)^2$	$\dfrac{ft^3}{lbmol}$	$atm\left(\dfrac{ft^3}{lbmol}\right)^2(°R)^{1/2}$	$\dfrac{ft^3}{lbmol}$
Air	345	0.586	5,409	0.4064
Butane (C_4H_{10})	3,509	1.862	98,349	1.2903
Carbon dioxide (CO_2)	926	0.686	21,972	0.4755
Carbon monoxide (CO)	372	0.632	5,832	0.4382
Methane (CH_4)	581	0.685	10,919	0.4751
Nitrogen (N_2)	346	0.618	5,280	0.4286
Oxygen (O_2)	349	0.509	5,896	0.3531
Propane (C_3H_8)	2,369	1.444	61,952	1.0006
Refrigerant 12	2,660	1.558	70,951	1.0796
Sulfur dioxide (SO_2)	1,738	0.910	49,032	0.6309
Water (H_2O)	1,400	0.488	48,418	0.3380

Source: Calculated from critical data.

2. Benedict–Webb–Rubin: Constants for pressure in atm, specific volume in $ft^3/lbmol$, and temperature in °R

Substance	a	A	b	B	c	C	α	γ
C_4H_{10}	7736.7	2587.6	10.26	1.9921	4.214×10^9	8.254×10^8	4.527	8.724
CO_2	562.3	702.4	1.850	0.7995	1.987×10^8	1.152×10^8	0.348	1.384
CO	150.6	344.1	0.675	0.8737	1.385×10^7	7.118×10^6	0.555	1.540
CH_4	203.0	476.0	0.867	0.6824	3.389×10^7	1.876×10^7	0.511	1.540
N_2	103.2	270.4	0.597	0.6526	9.700×10^6	6.700×10^6	0.523	1.360

Source: H. W. Cooper and J. C. Goldfrank, *Hydrocarbon Processing*, *46* (12): 141 (1967).

Table A-24E

TABLE A-25E

Thermochemical Properties of Selected Substances at 537°R and 1 atm

Substance	Formula	Molar Mass, M (lb/lbmol)	Enthalpy of Formation, $\bar{h}_f^\circ$ (Btu/lbmol)	Gibbs Function of Formation, $\bar{g}_f^\circ$ (Btu/lbmol)	Absolute Entropy, $\bar{s}^\circ$ (Btu/lbmol·°R)	Heating Values Higher, HHV (Btu/lb)	Heating Values Lower, LHV (Btu/lb)
Carbon	C(s)	12.01	0	0	1.36	14,100	14,100
Hydrogen	H_2(g)	2.016	0	0	31.19	61,000	51,610
Nitrogen	N_2(g)	28.01	0	0	45.74	—	—
Oxygen	O_2(g)	32.00	0	0	48.98	—	—
Carbon monoxide	CO(g)	28.01	−47,540	−59,010	47.27	—	—
Carbon dioxide	CO_2(g)	44.01	−169,300	−169,680	51.03	—	—
Water	H_2O(g)	18.02	−104,040	−98,350	45.08	—	—
Water	H_2O(l)	18.02	−122,970	−102,040	16.71	—	—
Hydrogen peroxide	H_2O_2(g)	34.02	−58,640	−45,430	55.60	—	—
Ammonia	NH_3(g)	17.03	−19,750	−7,140	45.97	—	—
Oxygen	O(g)	16.00	107,210	99,710	38.47	—	—
Hydrogen	H(g)	1.008	93,780	87,460	27.39	—	—
Nitrogen	N(g)	14.01	203,340	195,970	36.61	—	—
Hydroxyl	OH(g)	17.01	16,790	14,750	43.92	—	—
Methane	CH_4(g)	16.04	−32,210	−21,860	44.49	23,880	21,520
Acetylene	C_2H_2(g)	26.04	97,540	87,990	48.00	21,470	20,740
Ethylene	C_2H_4(g)	28.05	22,490	29,306	52.54	21,640	20,290
Ethane	C_2H_6(g)	30.07	−36,420	−14,150	54.85	22,320	20,430
Propylene	C_3H_6(g)	42.08	8,790	26,980	63.80	21,050	19,700
Propane	C_3H_8(g)	44.09	−44,680	−10,105	64.51	21,660	19,950
Butane	C_4H_{10}(g)	58.12	−54,270	−6,760	74.11	21,300	19,670
Pentane	C_5H_{12}(g)	72.15	−62,960	−3,530	83.21	21,090	19,510
Octane	C_8H_{18}(g)	114.22	−89,680	7,110	111.55	20,760	19,270
Octane	C_8H_{18}(l)	114.22	−107,530	2,840	86.23	20,610	19,110
Benzene	C_6H_6(g)	78.11	35,680	55,780	64.34	18,180	17,460
Methanol	CH_3OH(g)	32.04	−86,540	−69,700	57.29	10,260	9,080
Methanol	CH_3OH(l)	32.04	−102,670	−71,570	30.30	9,760	8,570
Ethanol	C_2H_5OH(g)	46.07	−101,230	−72,520	67.54	13,160	11,930
Ethanol	C_2H_5OH(l)	46.07	−119,470	−75,240	38.40	12,760	11,530

Source: Based on JANAF Thermochemical Tables, NSRDS-NBS-37, 1971; *Selected Values of Chemical Thermodynamic Properties,* NBS Tech. Note 270-3, 1968; and *API Research Project 44,* Carnegie Press, 1953. Heating values calculated.

Index to Figures and Charts

Figure A-1 Generalized compressibility chart, $p_R \leq 1.0$. *Source*: E. F. Obert, *Concepts of Thermodynamics*, McGraw-Hill, New York, 1960.

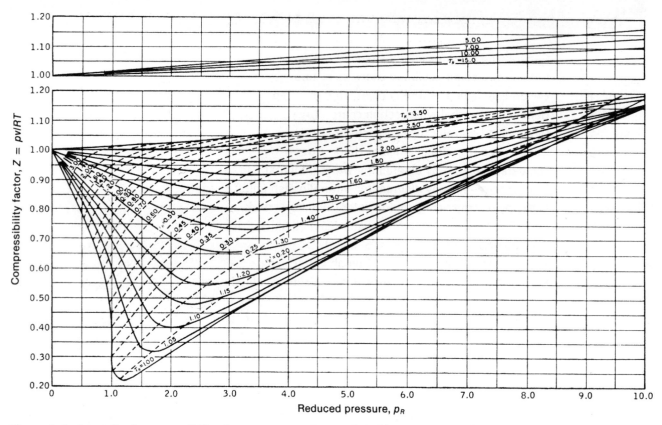

Figure A-2 Generalized compressibility chart, $p_R \leq 10.0$. *Source*: E. F. Obert, *Concepts of Thermodynamics*, McGraw-Hill, New York, 1960.

Figure A-3 Generalized compressibility chart, $10 \leq p_R \leq 40$. *Source*: E. F. Obert. *Concepts of Thermodynamics*, McGraw-Hill, New York, 1960.

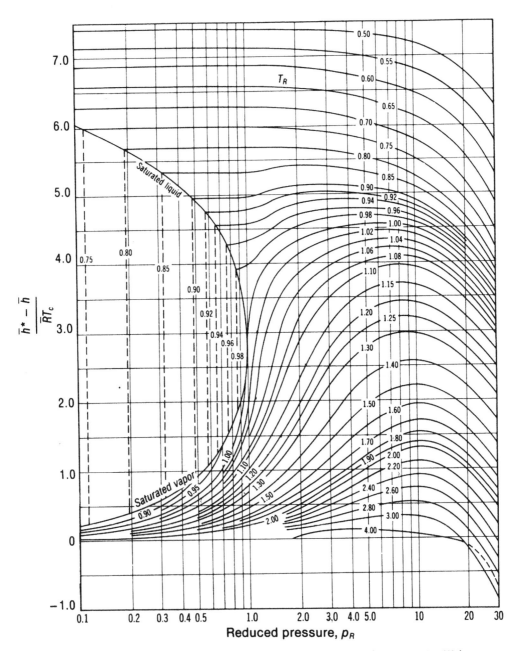

Figure A-4 Generalized enthalpy correction chart. *Source*: Adapted from G. J. Van Wylen and R. E. Sonntag, *Fundamentals of Classical Thermodynamics*, 3rd. ed., English/SI, Wiley, New York, 1986.

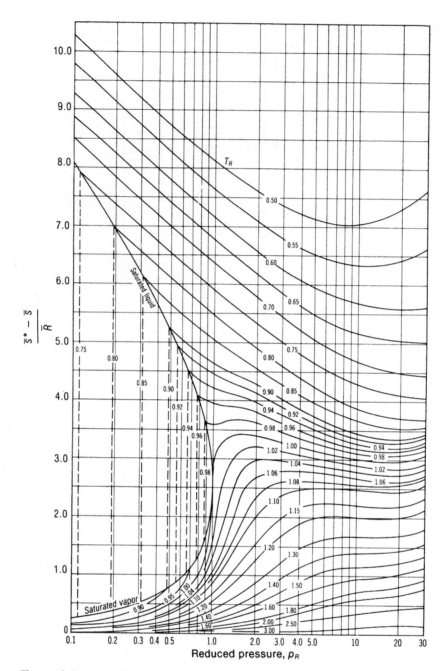

Figure A-5 Generalized entropy correction chart. *Source*: Adapted from G. J. Van Wylen and R. E. Sonntag, *Fundamentals of Classical Thermodynamics*, 3rd. ed., English/SI, Wiley, New York, 1986.

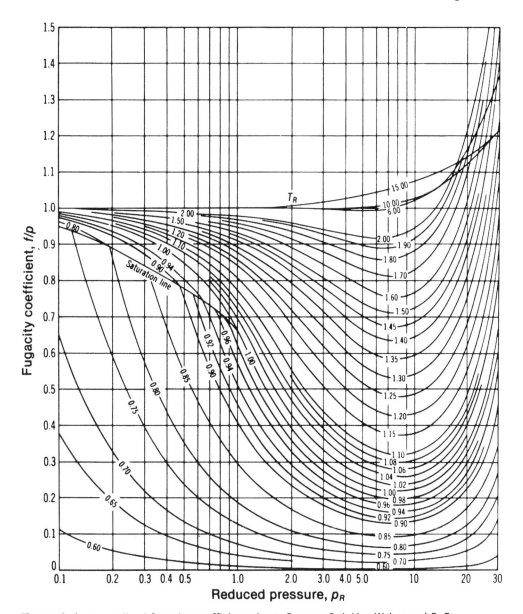

Figure A-6 Generalized fugacity coefficient chart. *Source*: G. J. Van Wylen and R. E. Sonntag, *Fundamentals of Classical Thermodynamics*, 3rd. ed., English/SI, Wiley, New York, 1986.

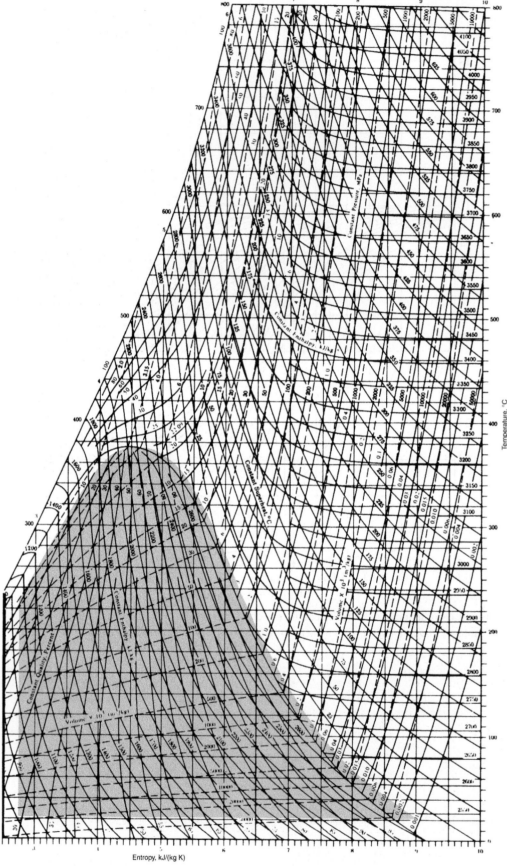

Figure A-7 Temperature–entropy diagram for water (SI units). *Source*: J. H. Keenan, F. G. Keyes, P. G. Hill, and J. G. Moore, *Steam Tables*, Wiley, New York, 1978.

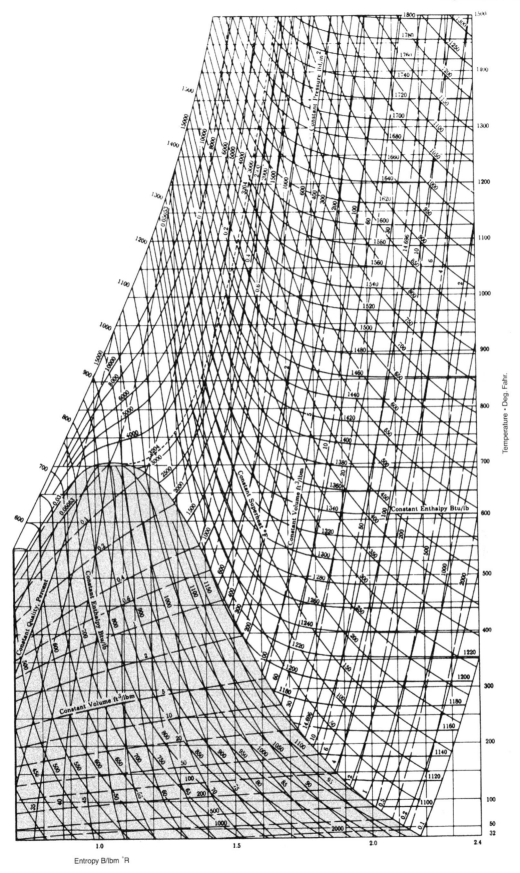

Figure A-7E Temperature–entropy diagram for water (English units). *Source*: J. H. Keenan, F. G. Keyes, P. G. Hill, and J. G. Moore, *Steam Tables*, Wiley, New York, 1969.

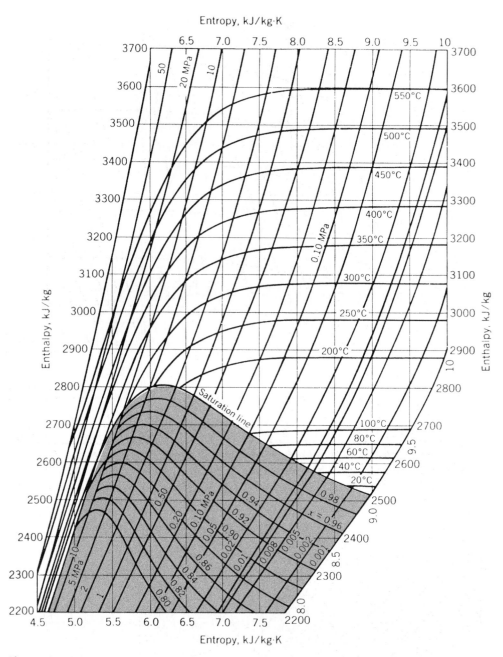

Figure A-8 Enthalpy–entropy diagram for water (SI units). *Source*: J. B. Jones and G. A. Hawkins, *Engineering Thermodynamics*, 2nd ed., Wiley, New York, 1986.

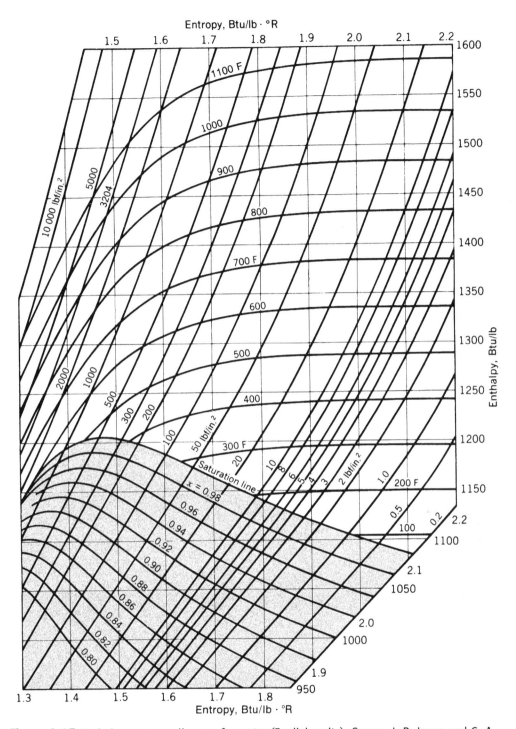

Figure A-8E Enthalpy–entropy diagram for water (English units). *Source*: J. B. Jones and G. A. Hawkins, *Engineering Thermodynamics*, 2nd ed., Wiley, New York, 1986.

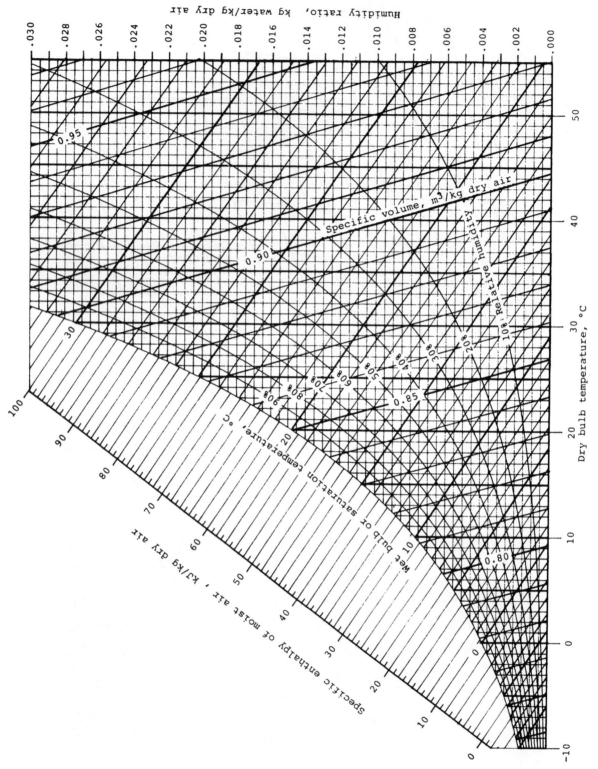

Figure A-9 Psychrometric chart for 1 atm (SI units). *Source*: Z. Zhang and M. B. Pate, "A Methodology for Implementing a Psychrometric Chart in a Computer Graphics System," *ASHRAE Transactions*, Vol. 94, Pt. 1, 1988.

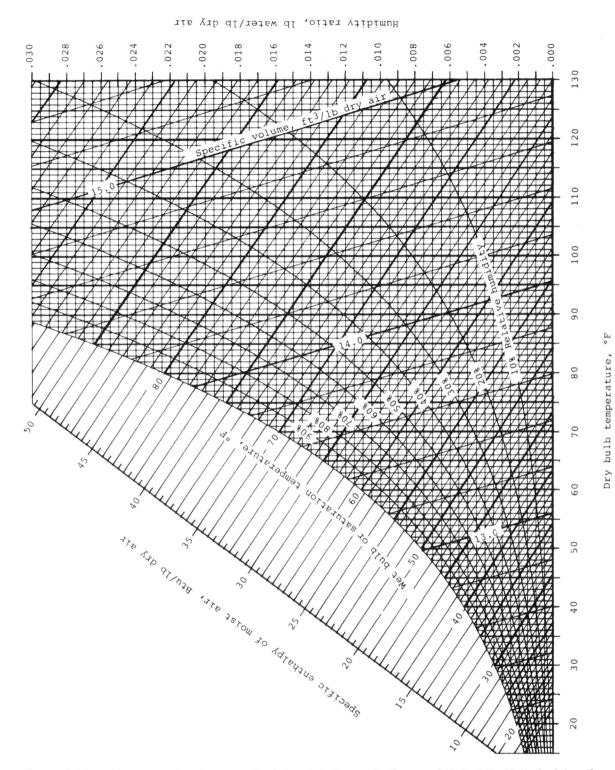

Figure A-9E Psychrometric chart for 1 atm (English units). *Source*: Z. Zhang and M. B. Pate, "A Methodology for Implementing a Psychrometric Chart in a Computer Graphics System," *ASHRAE Transactions*, Vol. 94, Pt. 1, 1988.

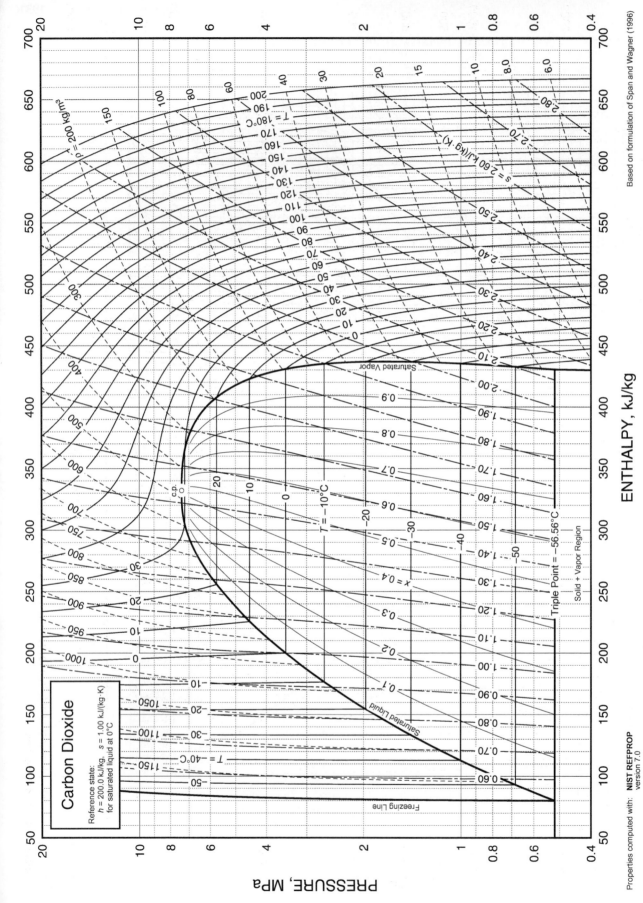

Properties computed with: **NIST REFPROP**
version 7.0

Based on formulation of Span and Wagner (1996)

Figure A-10 Pressure-enthalpy diagram for carbon dioxide (SI units). *Source:* ©ASHRAE, www.ashrae.org. 2009 ASHRAE Handbook of Fundamentals—Fundamentals.

ENTHALPY, kJ/kg

PRESSURE, MPa

Properties computed with: **NIST REFPROP** version 7.0

Based on formulation of Span and Wagner (1996)

Figure A-10E Pressure-enthalpy diagram for carbon dioxide (English units). *Source:* ©ASHRAE, www.ashrae.org. 2009 ASHRAE Handbook of Fundamentals—Fundamentals.

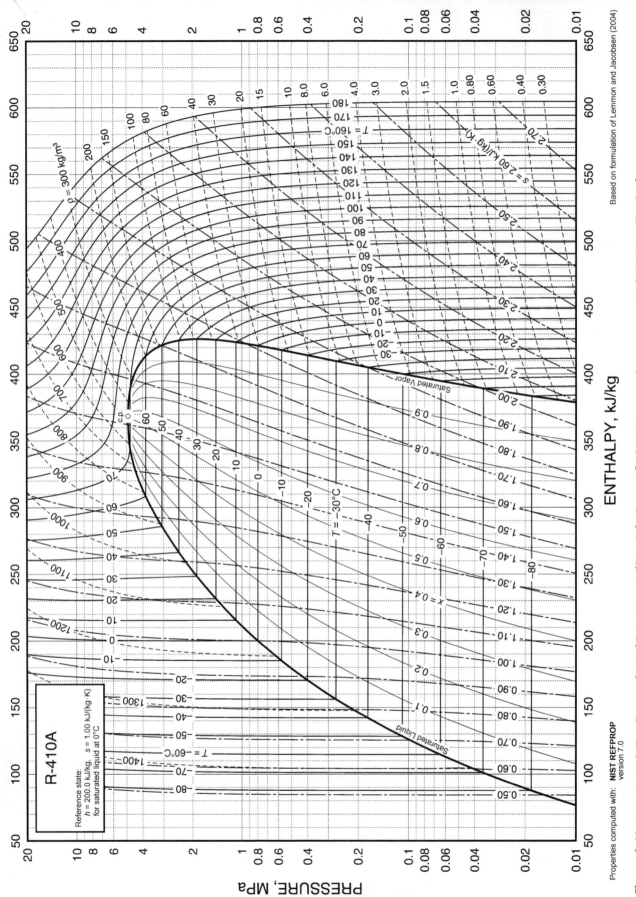

ENTHALPY, kJ/kg

Properties computed with: **NIST REFPROP** version 7.0

Based on formulation of Lemmon and Jacobsen (2004)

Figure A-11 Pressure-enthalpy diagram for Refrigerant 410A (SI units). *Source:* ©ASHRAE, www.ashrae.org. 2009 ASHRAE Handbook of Fundamentals—Fundamentals.

Properties computed with: **NIST REFPROP** version 7.0

Based on formulation of Lemmon and Jacobsen (2004)

Figure A-11E Pressure-enthalpy diagram for Refrigerant 410A (English units). *Source:* ©ASHRAE, www.ashrae.org. 2009 ASHRAE Handbook of Fundamentals—Fundamentals.

Conversion Factors

Mass and Density

1 kg	= 2.2046 lb
1 g/cm^3	= 10^3 kg/m^3
1 g/cm^3	= 62.428 lb/ft^3
1 lb	= 0.4536 kg
1 lb/ft^3	= 0.016018 g/cm^3
1 lb/ft^3	= 16.018 kg/m^3

Length

1 cm	= 0.3937 in.
1 m	= 3.2808 ft
1 in.	= 2.54 cm
1 ft	= 0.3048 m

Velocity

1 km/h	= 0.62137 mile/h
1 mile/h	= 1.6093 km/h

Volume

1 cm^3	= 0.061024 in.^{3}
1 m^3	= 35.315 ft^3
1 L	= 10^{-3} m^3
1 L	= 0.0353 ft^3
1 in.^{3}	= 16.387 cm^3
1 ft^3	= 0.028317 m^3
1 gal	= 0.13368 ft^3
1 gal	= 3.7854 × 10^{-3} m^3

Force

1 N	= 1 kg · m/s^2
1 N	= 0.22481 lbf
1 lbf	= 32.174 lb · ft/s^2
1 lbf	= 4.4482 N

Pressure

1 Pa	= 1 N/m^2
	= 1.4504 × 10^{-4} lbf/in.^{2}
1 bar	= 10^5 N/m^2
1 atm	= 1.01325 bar
1 lbf/in.^{2}	= 6894.8 Pa
1 lbf/in.^{2}	= 144 lbf/ft^2
1 atm	= 14.696 lbf/in.^{2}

Energy and Specific Energy

1 J	= 1 N · m = 0.73756 ft · lbf
1 kJ	= 737.56 ft · lbf
1 kJ	= 0.9478 Btu
1 kJ/kg	= 0.42992 Btu/lb
1 ft · lbf	= 1.35582 J
1 Btu	= 778.17 ft · lbf
1 Btu	= 1.0551 kJ
1 Btu/lb	= 2.326 kJ/kg
1 kcal	= 4.1868 kJ

Energy Transfer Rate

1 W	= 1 J/s = 3.413 Btu/h
1 kW	= 1.341 hp
1 Btu/h	= 0.293 W
1 hp	= 2545 Btu/h
1 hp	= 550 ft · lbf/s
1 hp	= 0.7457 kW

Specific Heat

1 kJ/kg · K	= 0.238846 Btu/lb · °R
1 kcal/kg · K	= 1 Btu/lb · °R
1 Btu/lb · °R	= 4.1868 kJ/kg · K

Others

1 ton of refrigeration = 200 Btu/min = 211 kJ/min
1 volt = 1 watt per ampere

Constants

Universal Gas Constant

$$\bar{R} = \begin{cases} 8.314 \text{ kJ/kmol} \cdot \text{K} \\ 1545 \text{ ft} \cdot \text{lbf/lbmol} \cdot °\text{R} \\ 1.986 \text{ Btu/lbmol} \cdot °\text{R} \end{cases}$$

Standard Acceleration of Gravity

$$g = \begin{cases} 9.80665 \text{ m/s}^2 \\ 32.174 \text{ ft/s}^2 \end{cases}$$

Standard Atmospheric Pressure

$$1 \text{ atm} = \begin{cases} 1.01325 \text{ bar} \\ 14.696 \text{ lbf/in.}^2 \\ 760 \text{ mm Hg} = 29.92 \text{ in. Hg} \end{cases}$$

Temperature Relations

$T(°\text{R}) = 1.8\, T(\text{K})$
$T(°\text{C}) = T(\text{K}) - 273.15$
$T(°\text{F}) = T(°\text{R}) - 459.67$

Symbols

a	acceleration, activity	mep	mean effective pressure
A	area	mf	mass fraction
AF	air–fuel ratio	n	number of moles, polytropic exponent
bwr	back work ratio	N	number of components in the phase rule
c	specific heat of an incompressible substance, velocity of sound	p	pressure
c	unit cost	p_{atm}	atmospheric pressure
$\dot{C}$	cost rate	p_i	pressure associated with mixture component i, partial pressure of i
C_aH_b	hydrocarbon fuel	p_r	relative pressure as used in Tables A-22
c_p	specific heat at constant pressure, $\partial h/\partial T)_p$	p_R	reduced pressure: p/p_c
c_v	specific heat at constant volume, $\partial u/\partial T)_v$	P	number of phases in the phase rule
c_{p0}	specific heat c_p at zero pressure	**P**	electric dipole moment per unit volume
e, E	energy per unit of mass, energy	pe, PE	potential energy per unit of mass, potential energy
e, E	exergy per unit of mass, exergy		
$e_f, \dot{E}_f$	specific flow exergy, flow exergy rate	$\dot{q}$	heat flux
$E_d, \dot{E}_d$	exergy destruction, exergy destruction rate	Q	heat transfer
$E_q, \dot{E}_q$	exergy transfer accompanying heat transfer, rate of exergy transfer accompanying heat transfer	$\dot{Q}$	heat transfer rate
		$\dot{Q}_x$	conduction rate
E_w	*exergy* transfer accompanying work	$\dot{Q}_c, \dot{Q}_e$	convection rate, thermal radiation rate
E	electric field strength	r	compression ratio
$\mathscr{E}$	electrical potential, electromotive force (emf)	r_c	cutoff ratio
f	fugacity	R	gas constant: $\overline{R}/M$, resultant force, electric resistance
$\overline{f}_i$	fugacity of component i in a mixture	$\overline{R}$	universal gas constant
F	degrees of freedom in the phase rule	s, S	entropy per unit of mass, entropy
F, F	force vector, force magnitude	$s°$	entropy function as used in Tables A-22, absolute entropy at the standard reference pressure as used in Table A-23
FA	fuel–air ratio		
g	acceleration of gravity		
g, G	Gibbs function per unit of mass, Gibbs function	t	time
$\overline{g}_f°$	Gibbs function of formation per mole at standard state	T	temperature
		T_R	reduced temperature: T/T_c
h, H	enthalpy per unit of mass, enthalpy	$\mathscr{T}$	torque
h	heat transfer coefficient	u, U	internal energy per unit of mass, internal energy
H	magnetic field strength	v, V	specific volume, volume
$\overline{h}_f°$	enthalpy of formation per mole at standard state	**V**, V	velocity vector, velocity magnitude
		v_r	relative volume as used in Tables A-22
$\overline{h}_{RP}$	enthalpy of combustion per mole	v'_R	pseudoreduced specific volume: $\overline{v}/(\overline{R}T_c/p_c)$
HHV	higher heating value	V_i	volume associated with mixture component i, partial volume of i
i	electric current		
k	specific heat ratio: c_p/c_v	W	work
k	Boltzmann constant	$\dot{W}$	rate of work, or power
K	equilibrium constant	x	quality, position
ke, KE	kinetic energy per unit of mass, kinetic energy	X	extensive property
l, L	length	y	mole fraction, mass flow rate ratio
LHV	lower heating value	z	elevation, position
m	mass	Z	compressibility factor, electric charge
$\dot{m}$	mass flow rate	$\dot{Z}$	cost rate of owning/operating
M	molecular weight, Mach number		
M	magnetic dipole moment per unit volume		

Greek Letters

α	isentropic compressibility
β	coefficient of performance for a refrigerator, volume expansivity
γ	coefficient of performance for a heat pump, activity coefficient
Δ	change = final minus initial
ε	exergetic (second law) efficiency, emissivity, extent of reaction
η	efficiency, effectiveness
θ	temperature
κ	thermal conductivity, isothermal compressibility
μ	chemical potential
μ_J	Joule–Thomson coefficient
ν	stoichiometric coefficient
ρ	density
$\sigma, \dot{\sigma}$	entropy production, rate of entropy production
σ	normal stress, Stefan–Boltzmann constant
Σ	summation
τ	surface tension
ϕ	relative humidity
ψ, Ψ	Helmholtz function per unit of mass, Helmholtz function
ω	humidity ratio (specific humidity), angular velocity

Subscripts

a	dry air
ad	adiabatic
as	adiabatic saturation
ave	average
b	boundary
c	property at the critical point, compressor, overall system
cv	control volume
cw	cooling water
C	cold reservoir, low temperature
db	dry bulb
e	state of a substance exiting a control volume
e	exergy reference environment
f	property of saturated liquid, temperature of surroundings, final value
F	fuel
fg	difference in property for saturated vapor and saturated liquid
g	property of saturated vapor
H	hot reservoir, high temperature
i	state of a substance entering a control volume, mixture component
i	initial value, property of saturated solid
I	irreversible
ig, if	difference in property for saturated vapor (saturated liquid) and saturated solid
isol	isolated
int rev	internally reversible
j	portion of the boundary, number of components present in a mixture
n	normal component
p	pump
ref	reference value or state
reg	regenerator
res	reservoir
P	products
R	reversible, reactants
s	isentropic
sat	saturated
surr	surroundings
t	turbine
tp	triple point
0	property at the dead state, property of the surroundings
o	stagnation property
v	vapor
w	water
wb	wet bulb
x	upstream of a normal shock
y	downstream of a normal shock
1,2,3	different states of a system, different locations in space

Superscripts

ch	chemical exergy
e	component of the exergy reference environment
–	bar over symbol denotes property on a molar basis (over X, V, H, S, U, Ψ, G, the bar denotes partial molal property)
.	dot over symbol denotes time rate
○	property at standard state or standard pressure
*	ideal gas, quantity corresponding to sonic velocity